ibvt-Schriftenreihe

Schriftenreihe des Institutes für Bioverfahrenstechnik
der Technischen Universität Braunschweig

Herausgegeben von Prof. Dr. Rainer Krull

Band 84

Cuvillier-Verlag
Göttingen, Deutschland

Herausgeber
Prof. Dr. Rainer Krull
Institut für Bioverfahrenstechnik
TU Braunschweig
Rebenring 56, 38106 Braunschweig
www.ibvt.de

Bibliographische Informationen der Deutschen Nationalbibliothek
Die Deutsche Nationalbibliothek verzeichnet diese Publikation in der Deutschen Nationalbibliographie; detaillierte bibliographische Daten sind im Internet über *http://dnb.d-nb.de* abrufbar.
1. Aufl. – Göttingen: Cuvillier, 2021

Nonnenstieg 8, 37075 Göttingen
Telefon: 0551-54724-0
Telefax: 0551-54724-21
www.cuvillier.de

1. Auflage, 2021
Gedruckt auf säurefreiem Papier

ISBN 978-3-7369-7350-3
eISBN 978-3-7369-6350-4
ISSN 1431-7230

Micro- and Macroparticle enhanced cultivation of filamentous *Lentzea aerocolonigenes* for increased rebeccamycin production

Von der Fakultät für Maschinenbau

der Technischen Universität Carolo-Wilhelmina zu Braunschweig

zur Erlangung der Würde

einer Doktor-Ingenieurin (Dr.-Ing.)

genehmigte Dissertation

von: M.Sc. Kathrin Schrinner

geboren in: Salzgitter

eingereicht am: 22.09.2020

mündliche Prüfung am: 11.12.2020

Vorsitz: Prof. Dr.-Ing. Arno Kwade

Gutachter: Prof. Dr. Rainer Krull

Prof. Dr. techn. Christoph Herwig

2020

Danksagung

Die vorliegende Arbeit ist während meiner Tätigkeit als wissenschaftliche Mitarbeiterin am Institut für Bioverfahrenstechnik an der Technischen Universität Braunschweig entstanden und wurde im Rahmen des Schwerpunktprogramms SPP 1934 DiSPBiotech „Dispersitäts-, Struktur- und Phasenänderungen von Proteinen und biologischen Agglomeraten in biotechnologischen Prozessen“ von der Deutschen Forschungsgemeinschaft gefördert.

Mein besonderer Dank gilt Prof. Rainer Krull für die Betreuung meiner Arbeit, die Unterstützung während dieser Zeit und die gewährten Freiheiten bei der Gestaltung des Projekts. Prof. Christoph Herwig danke ich für die Übernahme des Koreferats und Prof. Arno Kwade für die Übernahme des Prüfungsvorsitzes.

Bei meinem Projektpartner Marcel Schrader bedanke ich mich für die gute Zusammenarbeit und die stetige Diskussionsbereitschaft, besonders in schwierigen Phasen.

Meinen Studenten Jülide Gürler, Franziska Teubner, Friederike Schwarzer, Leon Klose, Henry Brauns, Jana Niebusch, Martina Graf, Kristin Althof, Nadine Wurzler, Paul Nowka, Valeria Anzorandía, Artem Kalinin, Anna Ebersbach und Annika Wilhelms danke ich für ihren Fleiß und ihr Engagement in ihren Abschlussarbeiten, Praktika oder als studentische Hilfskräfte.

Mein Dank gilt allen Mitgliedern des ibvt für die tolle Arbeitsatmosphäre. Besonders hervorzuheben sind hierbei Yvonne Göcke, Elena Kempf, Cord Hullmann und Detlev Rasch, die jederzeit hilfreich zur Seite standen, sowie meinem Bürokollegen Sebastian Tesche.

Den Mitgliedern des Schwerpunktprogramms DiSPBiotech und besonders der „Bioagglomerat-Gruppe“ danke ich für tolle Diskussionen, eine gute Zusammenarbeit sowie eine nette Atmosphäre bei den regelmäßigen Treffen. Sissy Bliatsiou, Philipp Waldherr und Stefan Schmideder möchte ich besonders für die produktiven Kooperationen danken.

Lukas Veiter von der TU Wien danke ich für die unkomplizierte Zusammenarbeit in unserem Kooperationsprojekt.

Zu guter Letzt möchte ich meiner Familie und meinen Freunden danken, die mich in der gesamten Zeit unterstützt haben. Ein besonderer Dank gilt dabei meinem Mann, der mich in allen Phasen der Arbeit ertragen und immer viel Verständnis aufgebracht hat.

Kathrin Schrinner (geb. Pommerehne) Braunschweig, Dezember 2020

Publications

Articles in Journals

Schrinner, K., Althof, K., Ebersbach, A.T., Grosch, J.-H., Krull, R. (2020) XAD resins for increased rebeccamycin productivity in cultivations of *Lentzea aerocolonigenes*. Chem Ing Tech. https://doi.org/10.1002/cite.202000131

Böl, M., **Schrinner, K.**, Tesche, S., Krull, R. (2020) Challenges of influencing cellular morphology by morphology engineering techniques and mechanical induced stress on filamentous pellet systems – a critical review. Eng Life Sci. https://doi.org/10.1002/elsc.202000060

Bliatsiou, C., **Schrinner, K.**, Waldherr, P., Tesche, S., Böhm, L., Kraume, M., Krull, R. (2020) Rheological characteristics of filamentous cultivation broths and suitable model fluids. Biochem Eng J 163:107746. https://doi.org/10.1016/j.bej.2020.107746

Schrinner, K., Veiter, L., Schmideder, S., Doppler, P., Schrader, M., Münch, N., Althof, K., Kwade, A., Briesen, H., Herwig, C., Krull, R. (2020) Morphological and physiological characterization of filamentous *Lentzea aerocolonigenes*: Comparison of biopellets by microscopy and flow cytometry. PLoS ONE 15(6): e0234125. https://doi.org/10.1371/journal.pone.0234125

Schrader, M., **Pommerehne, K.**, Wolf, S., Finke, B., Schilde, C., Kampen, I., Lichtenegger, T., Krull, R., Kwade, A. (2019) Design of a CFD-DEM-based method for mechanical stress calculation and its application to glass bead-enhanced cultivations of filamentous *Lentzea aerocolonigenes*. Biochem Eng J 148, 116-130, https://doi.org/10.1016/j.bej.2019.04.014

Pommerehne, K., Walisko, J., Ebersbach, A., Krull, R. (2019) The antitumor antibiotic rebeccamycin – Challenges and advanced approaches in production processes. Appl Microbiol Biotech 103, 3627-3636, https://doi.org/10.1007/s00253-019-09741-y

Oral Presentations

Schrinner, K., Schrader, M., Niebusch, J., Wurzler, N., Kwade, A., Krull, R. (2019) Particle based cultivation of Lentzea aerocolonigenes in membrane aerated stirred bioreactors for increased rebeccamycin production. *3rd International Symposium on Pharmaceutical Engineering Research – SphERe*, Braunschweig. https://doi.org/10.24355/dbbs.084-202001221448-0

Schrinner, K., Wurzler, N., Schrader, M., Kwade, A., Krull, R. (2019) Mechanical stress during scale up to membrane aerated stirred bioreactors for rebeccamycin production in filamentous Lentzea aerocolonigenes. *12th European Congress of Chemical Engineering & 5th European Congress of Applied Biotechnology*, Florence, Italy.

Schrader, M., **Pommerehne, K.**, Krull, R., Kwade, A. (2019) Experimentelle und simulative Untersuchungen partikelinduzierter mechanischer Beanspruchungen auf filamentöse Bakterien in Schüttelkolben als Basis für den ScaleUp. *22. Köthener Rührer-Kolloquium, Verfahrenstechnische Charakterisierungsmethoden von Mischprozessen*, Köthen.

Böhm, L., Bliatsiou, C., Panckow, R., Waldherr, P., Schestkowa, H., Oechsle, A., Drusch, S., Kraume, M., **Pommerehne, K.**, Kampen, I., Krull, R., Kwade, A., Schmideder, S., Nowotny, P., Sedlmeier, S., Briesen, H., Kulozik,U., Weuster-Botz, D., Seidel, J., Lode, A., Steingroewer, J., Barros Groß, M., Radel, B., Kind, M., Nirschl, H., Schmidt, M.-P., Hirsch,S., Maaß, S., Veiter, L., Herwig, C. (2018) Partikelcharakterisierung und -messtechniken in lebensmittel- und biotechnologischen Systemen. *ProcessNet-Jahrestagung und 33. DECHEMA-Jahrestagung der Biotechnologen*, Aachen. https://doi.org/10.1002/cite.201871204

Pommerehne, K., Schrader, M., Bliatsiou, C., Schmideder, S., Böhm, L., Briesen, H., Kraume, M., Kwade, A., Krull R. (2018) Experimental and numerical investigations on cultivations of filamentous microorganisms towards a better understanding and process control. *ProcessNet-Jahrestagung und 33. DECHEMA-Jahrestagung der Biotechnologen*, Aachen. https://doi.org/10.1002/cite.201855449

Pommerehne, K., Schrader, M., Kwade, A., Krull, R. (2018) Influence of glass particles on heterogeneous morphology and productivity of the filamentous bacterium Lechevalieria aerocolonigenes. Dechema-Himmelfahrtstagung *Heterogeneities - A key for understanding and upscaling of bioprocesses in up- and downstream*, Magdeburg.

Pommerehne, K., Wilhelms, A., Krull, R. (2017) Particle-induced effects on antibiotic production of *Lechevalieria aerocolonigenes*. Laboratoire d'Ingénierie des Systèmes Biologiques et des Procédés (LISBP), INRA/INSA, Toulouse, France.

Poster Presentations

Schrader, M., **Schrinner, K.**, Klose, L., Brauns, H., Kampen, I., Krull, R., Kwade, A. (2020) Morphology engineering of filamentous Lentzea aerocolonigenes with chemical modified micro-particles. *10. ProcessNet-Jahrestagung und 34. DECHEMA-Jahrestagung der Biotechnologen*.

Schrader, M., **Pommerehne, K.**, Krull, R., Kwade, A. (2019) Quantification of particle-induced mechanical stress on filamentous microorganisms via CFD-DEM simulations. *DEM 8 – 8th International Conference on Discrete Element Methods*, Enschede, The Netherlands.

Schrader, M., **Pommerehne, K.**, Schilde, C., Pirker, S., Krull, R., Kwade, A. (2018) Coupled CFD-DEM-simulation and experimental validation of morphology and productivity by particle-induced mechanical stress on the filamentous system of *Lechevalieria aerocolonigenes*. *ProcessNet-Jahrestagung und 33. DECHEMA-Jahrestagung der Biotechnologen*, Aachen. https://doi.org/10.1002/cite.201855451

Bliatsiou, C., **Pommerehne, K.**, Tesche, S., Waldherr, P., Böhm, L., Krull, R., Kraume, M. (2018) Rheological characteristics of filamentous cultivation broths and suitable model systems. Dechema-Himmelfahrtstagung *Heterogeneities - A key for understanding and upscaling of bioprocesses in up- and downstream,* Magdeburg.

Krull, R., **Pommerehne, K.**, Walisko, J., Tesche, S. (2017) Morphology engineering of stress sensitive filamentous microorganisms. *10th World Congress of Chemical Engineering (WCCE 10),* Barcelona, Spain.

Pommerehne, K., Wilhelms, A., Walisko, J., Krull, R. (2017) Increase of Rebeccamycin production with *Lechevalieria aerocolonigenes* using particle-based morphology engineering. *2nd Symposium on Pharmaceutical Engineering Research – SphERe*, Braunschweig.

Pommerehne, K., Wilhelms, A., Walsiko, J., Krull, R. (2017) Morphology engineering of *Lechevalieria aerocolonigenes* for the production of the antibiotic Rebeccamycin. Dechema-Himmelfahrtstagung *Models for Developing and Optimising Biotech Production*, Neu-Ulm.

Students' theses

Gürler, J. (2020) Einfluss von oberflächenmodifizierten Glasmikropartikeln auf die Produktbildung und die Morphologie in Kultivierungen von *Lentzea aerocolonigenes*. Bachelor thesis (142) Institute of Biochemical Engineering/Institute for Particle Technology, TU Braunschweig.

Teubner, F. (2019) Steigerung der Rebeccamycinausbeute durch verschiedene Zellaufschlussmethoden. Forschungspraktikum, Institute of Biochemical Engineering, TU Braunschweig.

Schwarzer, F. A. (2019) Einfluss von Inokulumsvolumen und Lecithin auf die Produktion von Rebeccamycin in *Lentzea aerocolonigenes*. Bachelor thesis (139) Institute of Biochemical Engineering, TU Braunschweig.

Klose, L. (2019) Untersuchungen zur Funktionalisierung von Mikropartikeln und deren Einfluss auf die Rebeccamycin-Produktion in Kultivierungen von *Lentzea aerocolonigenes*. Bachelor thesis (134) Institute of Biochemical Engineering/Institute for Particle Technology, TU Braunschweig.

Brauns, H. (2019) Herstellung und Anwendung von oberflächenmodifizierten Mikropartikeln in Kultivierungen von *Lentzea aerocolonigenes*. Bachelor thesis (133) Institute of Biochemical Engineering/Institute for Particle Technology, TU Braunschweig.

Niebusch, J. (2019) Glaspartikel- und rührerinduzierte Beanspruchung von *Lentzea aerocolonigenes* im membranbegasten Bioreaktor zur Verbesserung der Produktivität. Bachelor thesis (132) Institute of Biochemical Engineering, TU Braunschweig.

Althof, K. (2019) Glas- und Adsorberpartikel-induzierte Verbesserung der Rebeccamycinproduktion in *Lentzea aerocolonigenes*. Master thesis (79) Institute of Biochemical Engineering, TU Braunschweig.

Wurzler, N. (2019) Charakterisierung der Rührer-induzierten Beanspruchung des filamentösen Actinomyceten *Lentzea aerocolonigenes*. Master thesis (78) Institute of Biochemical Engineering, TU Braunschweig.

Nowka, P. (2018) Partikelbasierte mechanische Beanspruchung von *Lechevalieria aercolonigenes* zur Verbesserung der Produktivität. Master thesis (76) Institute of Biochemical Engineering, TU Braunschweig.

Kalinin, A. (2018) Optimierung der Animpfstrategie von *Lechevalieria aerocolonigenes* für erhöhte Vergleichbarkeit der Kultivierungen. Forschungspraktikum, Institute of Biochemical Engineering, TU Braunschweig.

Ebersbach, A. T. (2018) Productivity increase and process integrated isolation of rebeccamycin from *Lechevalieria aerocolonigenes* by adsorbing particles. Master thesis (61) Institute of Biochemical Engineering, TU Braunschweig.

Wilhelms, A. (2017) Partikelinduzierte Kultivierung von *Lechevalieria aerocolonigenes* zur optimierten Rebeccamycin-Produktion. Bachelor thesis (110) Institute of Biochemical Engineering, TU Braunschweig.

Kurzfassung

Zwei Drittel der derzeit bekannten Antibiotika werden von Actinomyceten produziert, was ihre Kultvierung erstrebenswert macht. Der filamentöse Actinomycet *Lentzea aerocolonigenes* produziert das antitumoral und antibiotisch wirkende Rebeccamycin. Die komplexe Morphologie der Actinomyceten besitzt allerdings einige kultivierungsbezogene Herausforderungen, die häufig mit geringen Produkttitern einhergehen. In letzter Zeit wurden durch den Zusatz von Partikeln zur Kultivierung von *L. aerocolonigenes* die bislang geringen Rebeccamycintiter deutlich erhöht. Daher sollte in dieser Arbeit der Zusatz von Mikro-, Makro- und Adsorberpartikeln zu *L. aerocolonigenes* genauer untersucht werden. Zusätzlich wurde eine Maßstabsvergrößerung in einen blasenfrei begasten Bioreaktor durchgeführt.

In Schüttelkolbenkultivierungen wurde durch Zusatz von Glasmikropartikeln (x_{50} = 7,9 µm, 10 g L^{-1}) der Rebeccamycintiter bis zu 3,6-fach gegenüber einer unsupplementierten Kontrollkultivierung erhöht. Dabei konnte über Pelletschnitte eine Einlagerung der Mikropartikel in die Pellets beobachtet werden. Mit verschiedenen Carbonsäure-Modifizierungen der Partikeloberfläche wurden teilweise höhere Rebeccamycintiter im Vergleich zu einer unsupplementierten Kontrollkultivierung erzielt. Modifizierte Mikropartikel, welche mit Hexansäure (HDA), 3-Mercaptopropionsäure (MPA) oder 3-Phenylpropionsäure (PPA) funktionalisiert wurden, erzeugten spezielle Einlagerungsmuster in den Pellets. Durch diese Partikeleinlagerung weist die Pelletmorphologie insgesamt eine lockerere Struktur auf, was zu einer verbesserten Nährstoff- und Sauerstoffversorgung in das Pelletinnere führt. Diese Beobachtung wird auch durch eine gesteigerte Pelletviabilität unter Mikropartikelzugabe unterstützt.

Der Zusatz von größeren (Glas-) Makropartikeln (ø = 0.2 - 2.1 mm, 100 g L^{-1}) steigerte ebenfalls den Rebeccamycintiter gegenüber der unsupplementierten Kontrollkultivierung, was auf die mechanische Beanspruchung der Biopellets durch die Glaskugeln zurückzuführen ist. Um eine maßgeschneiderte mechanischen Beanspruchung für eine optimale Rebeccamycinproduktion einzustellen, wurde der Einfluss verschiedener Partikeldurchmesser, -konzentrationen und -dichten während der Kultivierung untersucht. Bei der Kombination von Glaspartikeln (ø = 969 µm, 100 g L^{-1}) mit 5 g L^{-1} Sojalecithin wurden 388 mg L^{-1} Rebeccamycin produziert, was eine der höchsten jemals erzielten finalen Rebeccamycinkonzentrationen darstellt.

Für die Maßstabsvergrößerung der Kultivierung wurde ein blasenfreies Membranbegasungs-system entwickelt, um das Auftreten von Schaum zu reduzieren und die mechanische Beanspruchung definierter in das Kultivierungssystem einzutragen. An die schlauchförmige Membran kann zusätzlich erhöhter Druck für einen erhöhten Sauerstofftransfer angelegt werden. In ersten Kultivierungen wurden bis zu 18 mg L^{-1} an Rebeccamycin hergestellt. Dies entspricht in etwa den Konzentrationen die in Schüttelkolbenkultivierungen ohne Zusätze erreicht werden.

XAD Adsorberpartikel sollten zur Aufnahme von Rebeccamycin durch Adsorption auf die Adsorberharze und eine dadurch erleichterte Aufarbeitung eingesetzt werden. Allerdings bewirkte der Zusatz bestimmter XAD-Partikel eine Steigerung der Rebeccamycintiter. Dies wurde vermutlich durch die Adsorption der Vorgängersubstanz Tryptophan auf die XAD-Partikel und eine anschließende direkte Übertragung auf den Mikroorganismus ermöglicht. Durch die Kombination von Adsorber- und Glaspartikeln (ø = 969 µm) konnte eine weitere Steigerung des Rebeccamycintiters erzeugt werden. Die unterschiedlichen Wirkmechanismen der beiden Partikeltypen, Adsorptionseffekte und mechanische Beanspruchung, beeinflussen sich nicht negativ und können daher beide zur Produktionssteigerung beitragen.

Abstract

Two-thirds of the currently known antibiotics are naturally produced by actinomycetes making their cultivation quite worthwhile. The filamentous actinomycete *Lentzea aerocolonigenes* produces the antitumor antibiotic rebeccamycin. However, the complex morphology of actinomycetes leads to challenges during the cultivation often accompanied by low product titers. In the recent past, the to date low rebeccamycin titers were increased by particle addition to cultivations of *L. aerocolonigenes*. In this thesis the addition of micro-, macro- and adsorbent particles to cultivations of *L. aerocolonigenes* were investigated in more detail. Furthermore, the scale-up to a bubble-free bioreactor was conducted.

The addition of glass microparticles (x_{50} = 7.9 µm, 10 g L^{-1}) to shake flask cultivations increased the rebeccamycin titer up to 3.6-fold compared to an unsupplemented approach. Pellet slices showed the incorporation of microparticles. With different surface modifications of the microparticles, caused by the functionalization with carboxylic acids, the rebeccamycin concentration was increased compared to an unsupplemented control in some cases. The modifications with hexanedioic acid (HDA), 3-mercaptopropionic acid (MPA) or 3-phenylpropanoic acid (PPA) caused specific incorporation patterns of the microparticles. The incorporation is likely influencing the pellet morphology to create more space between the hyphae allowing an increased nutrient and oxygen supply in the pellet core which is supported by an increased viability of pellets grown with microparticle addition.

An increased rebeccamycin concentration compared to an unsupplemented approach by the addition of larger (glass) macroparticles (ø = 0.2 - 2.1 mm, 100 g L^{-1}) is likely caused by the induction of mechanical stress on the biopellets. A tailor-made mechanical stress adjustment for an optimal rebeccamycin production is necessary. Hence, different particle diameters, particle concentrations and particle densities were investigated for their effects. The additional supplementation of 5 g L^{-1} soy lecithin and glass beads (ø = 969 µm, 100 g L^{-1}) resulted in a rebeccamycin titer of 388 mg L^{-1}, one of the highest rebeccamycin titers ever achieved.

For the scale-up of *L. aerocolonigenes* cultivations a bubble-free membrane aeration system was developed for reduced foaming and a more defined input of mechanical stress into the cultivation system. The tubular membrane aeration system can additionally be pressurized to increase the oxygen transfer. First cultivations of *L. aerocolonigenes* successfully provided 18 mg L^{-1} rebeccamycin, a concentration similar to that of unsupplemented shake flask cultivations.

The idea of the addition of XAD adsorbent resins in particle form to *L. aerocolonigenes* was a facilitated rebeccamycin recovery by adsorption to the resins. However, the XAD particles increased the rebeccamycin titer which was likely to be caused by the adsorption of the rebeccamycin precursor tryptophan to the resins which in turn directly transferred the tryptophan to the microorganism. The combination of adsorbent resins with glass particles (ø = 969 µm) further increased the rebeccamycin titer. Due to different mechanisms, adsorbent effects and mechanical stress, both particle types contribute to a rebeccamycin increase.

Table of contents

1 Introduction and aim of the thesis

The process of natural product formation in filamentous eukaryotic and prokaryotic microorganisms is popular and widely applied for research and industrial purposes. Cultivation of these microorganisms is challenging but also quite worthwhile because of their products. Regarding the pharmaceutical sector actinomycetes are of major importance for production. 60 % of the known secondary metabolites showing biological activity were isolated from actinomycetes in the year 2000 (Kieser et al. 2000). Many actinomycetes exhibit a fungal-like filamentous morphology which is one of the most challenging factors in the cultivation of filamentous microorganisms. These microorganisms are able to grow in different morphological forms ranging from dispersed mycelium to dense pellets (Manteca et al. 2008).

One example for an interesting secondary metabolite is rebeccamycin. Rebeccamycin is an antibiotic and antitumor substance. However, the antitumor effect is of primary interest. Rebeccamycin acts as a topoisomerase inhibitor and thereby interferes with DNA replication. Its analogues, that are easily derived from rebeccamycin, show a similar biological activity (Bush et al. 1987; Long et al. 2002). Becatecarin, an analogue with an increased water solubility, was already part of promising phase I and II clinical trials. In these trials becatecarin was tested for the treatment of e.g. metastatic colorectal cancer, refractory breast cancer or small cell lung cancer (Goel et al. 2003; Burstein et al. 2007; Schwandt et al. 2012). However, for the treatment of cancer patients large amounts of active substance are required. For instance, Schwandt et al. (2012) used a total of 4200 mg m^{-2} becatecarin per patient for the treatment of small cell lung cancer in their study. Regarding rebeccamycin titers described in literature mostly concentrations of under 100 mg L^{-1} are achieved. Moreover, the production is mostly conducted in small scales of around 100 mL (Pommerehne et al. 2019). Hence, research on the provision of larger amounts of rebeccamycin is necessary to allow a sufficient supply for a potential cancer treatment.

Lentzea aerocolonigenes is a filamentous actinobacterium that naturally produces rebeccamycin as one of several secondary metabolites. Due to the molecular structure of rebeccamycin, the biotechnological production potentially entails lower costs for production than a chemical synthesis (Burkart 2003), making the production in the original host *L. aerocolonigenes* more interesting. However, the manifold challenges during the cultivation of filamentous microorganisms have to be considered. One complex aspect is the morphology of these microorganisms. The extreme morphological characteristics are freely dispersed mycelium or dense pellets with various intermediate forms (Manteca et al. 2008). Mycelial structures or hairy pellets lead to an increased viscosity of the cultivation broth resulting in a reduced mass transfer which should be compensated with an increased power input (Wucherpfennig et al. 2010). However, dense and large pellets can lead to nutrient and oxygen limitations in the pellet core (Hille et al. 2005; Bizukojc and Gonciarz 2015). The most favorable morphology for optimal production yields is depending on the microorganism and the desired product (Whitaker 1992; Walisko et al. 2015).

A wide range of parameters is influencing the cell morphology during cultivation. For example, factors like the medium composition, pH value, osmolality, addition of microparticles, inoculation volume and method of inoculation (e.g., inoculation with spores or vegetative mycelial biomass from a pre-culture) can have a great impact. Moreover, operating conditions of the cultivation process, such as stirrer speed, stirrer geometry and aeration rate or type of aeration can influence growth by inducing (hydro)mechanical stress into the cultivation (Oncu et al. 2007; Papagianni 2004; Wucherpfennig et al. 2010; Walisko et al. 2015).

For the production of rebeccamycin in *L. aerocolonigenes*, Walisko (2017) considered some of these parameters in more detail. Especially the addition of (micro-)particles proved as a successful method for the enhancement of rebeccamycin titers. Walisko et al. (2017) used the addition of talc microparticles (10 g L^{-1}) to cultivations of *L. aerocolonigenes* to increase the rebeccamycin titer to around 120 mg L^{-1} compared to an unsupplemented cultivation with around 40 mg L^{-1}. Microscopic images indicated the entanglement of the talc microparticles in the pellets outer hyphal area. When talc microparticles with different hydrophilic and hydrophobic surface modifications (ø < 30 µm) were applied in cultivations, both increased and decreased rebeccamycin titers compared to an unsupplemented approach or with untreated talc particles were observed. Hence, Walisko et al. (2017) proposed physicochemical surface effects to play an important role in the mechanism for the increased product concentration. The addition of microparticles has also been successfully applied with several other filamentous microorganisms (Kaup et al. 2008; Driouch et al. 2011; Antecka et al. 2016a; Karahalil et al. 2019).

Furthermore, Walisko (2017) added larger glass particles (ø = 0.5 – 5 mm, mostly 80 g L^{-1}) to cultivations of *L. aerocolonigenes*. With 0.5 mm glass beads a rebeccamycin concentration of almost 120 mg L^{-1} compared to an unsupplemented approach with only 6 mg L^{-1} was achieved. With larger glass beads the rebeccamycin concentration was decreased, the pellet size decreased and with 3 and 5 mm glass beads only mycelial structures have been observed. In this case the mechanical stress induced by the glass beads were proposed to be responsible (Walisko 2017). Although the micro- and macroparticle addition proved to be beneficial in regard to rebeccamycin titers, the mechanisms behind these product increasing methods are not yet fully understood.

Due to the rather low rebeccamycin titers described in literature compared to the necessary amounts for cancer treatment, one objective of this thesis was to increase the achievable rebeccamycin concentrations in cultivations of *L. aerocolonigenes*. Different, mostly particle-based, cultivation methods were applied to achieve this aim. Moreover, a better understanding of the mechanisms responsible for the increased rebeccamycin titers with different cultivation methods was intended. In more detail, the aims of this thesis can be described as follows:

- The further investigation of the microparticle addition to cultivations of *L. aerocolonigenes* considering microparticle incorporation and physicochemical effects of surface modified microparticles,
- the characterization of mechanical stress effects of larger macroparticles in cultivations of *L. aerocolonigenes* by the variation of particle size, particle concentration and particle density,
- the development and application of a bubble-free aerated stirred bioreactor for the production of rebeccamycin in a larger scale under defined shear conditions,
- the particle induced rebeccamycin recovery for a facilitated downstream processing of rebeccamycin by addition of adsorbent resins in the shape of particles, and
- the consideration of the interdependency of the cell morphology of *L. aerocolonigenes* and production of rebeccamycin with the aforementioned methods applied.

2 Theoretical background

2.1 *Lentzea aerocolonigenes* and rebeccamycin

Lentzea aerocolonigenes DSM 44217 is an actinomycete originally isolated from a Panamanian soil sample and is obligately aerobic (Bush et al. 1987; Nettleton, Jr. et al. 1985). It is a Gram-positive bacterium (Labeda et al. 2001) with a high guanine-cytosine content (Labeda 1986). A characteristic of actinomycetes like *L. aerocolonigenes* is the complex morphology, in which hyphae can grow to form mycelium or pellets, and the formation of spores. The hyphae appear in a slightly yellowish color during growth (Bush et al. 1987). The microorganism was reclassified several times starting from *Streptomyces* (Shinobu and Kawato 1960) over *Nocardia* (Pridham 1970) and *Saccharothrix* (Labeda 1986) to *Lechevalieria* (Labeda et al. 2001) and finally *Lentzea aerocolonigenes* (Nouioui et al. 2018).

L. aerocolonigenes is a natural producer of various secondary metabolites. However, the most interesting natural product is rebeccamycin (1,11-dichloro-12-(4-O-methyl-beta-Dglucopyranosyl)-12,13-dihydro-5H-indolo[2,3-a]pyrrolo[3,4-c]carbazole-5,7(6H)-dione). It is a yellow crystalline and hydrophobic substance belonging to the indolocarbazoles with the indolocarbazole framework and two chlorine as well as one methylglucose substituents (**Figure 2.1 a**) (Nettleton, Jr. et al. 1985; Bush et al. 1987; Kaneko et al. 1985).

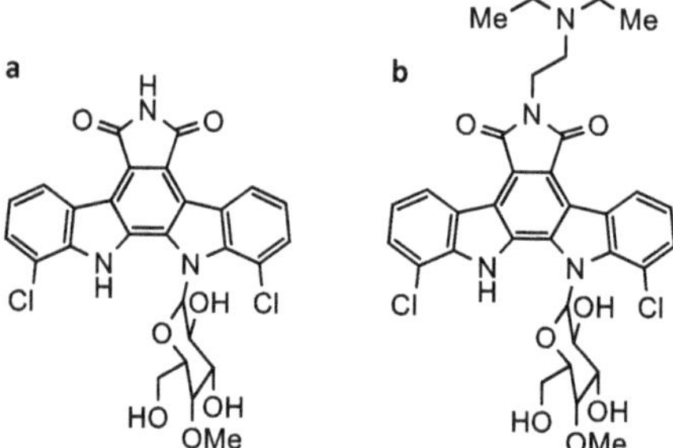

Figure 2.1: Chemical structures of the natural product (a) rebeccamycin and its analogue (b) becatecarin.

Rebeccamycin is poorly soluble in water (< 1 µg mL^{-1}) but is highly soluble in organic solvents (e.g. tetrahydrofuran and dimethylsulfoxide) (Bush et al. 1987; Nettleton, Jr. et al. 1985; Kaneko et al. 1990). It gained more interest than other natural products of *L. aerocolonigenes* because of its antibacterial effects against certain Gram-positive bacteria and additional antitumor effects, which makes it pharmaceutically relevant (Bush et al. 1987; Nettleton, Jr. et al. 1985).

The poor water solubility of rebeccamycin is critical for the use as a drug in the human body, which is why analogues with increased solubility in water were developed. Becatecarin (**Figure 2.1 b**) (1,11-Dichloro-6-[2-(diethylamino)ethyl]-12,13-dihydro-12-(4-O-methyl-beta-D-glucopyranosyl)-5H-indolo[2,3-a]pyrrolo[3,4-c]carbazole-5,7(6H)-dione) is a promising analogue with increased water solubility. It can be derived from rebeccamycin by adding an aminoalkyl group. The antibacterial and antitumor activity is maintained during the structural modification (Kaneko et al. 1990).

2.1.1 Medical applications

Rebeccamycin acts antibacterially against the Gram-positive bacteria *Staphylococcus aureus* and *Streptococcus faecalis*. Although this might already be interesting for a medical application, the antitumor properties of rebeccamycin are paramount. It is intercalating in the DNA and causes double-strand breaks as well as inhibits topoisomerase I (Bush et al. 1987; Facompré et al. 2002). The analogue becatecarin may even additionally inhibit topoisomerase II (Long et al. 2002). However, due to the low water solubility of rebeccamycin, mostly analogues were further investigated in clinical trials. Several phase I and II clinical trials with becatecarin with regard to small cell lung cancer, breast cancer, metastatic colorectal cancer and metastatic renal cell cancer were conducted (Nock et al. 2011; Burstein et al. 2007; Hussain et al. 2003; Schwandt et al. 2012; Goel et al. 2003). Furthermore, the effects on solid tumors in children was studied in clinical trials (Lagmay et al. 2016; Sánchez et al. 2006a; Langevin et al. 2008). The applied doses of becatecarin to children with solid tumors in the study of Langevin et al. (2008) was 650 mg m^{-2} every 21 days for maximal 16 cycles. However, in certain cases an increased dose of 780 mg m^{-2} per cycle was given to the patients. For the treatment of small cell lung cancer in adults 140 mg m^{-2} becatecarin each day for five days with a repetition every three weeks for maximal 6 cycles was applied in the study of Schwandt et al. (2012).

The applied doses in these clinical trials are rather high compared to rebeccamycin titers from biotechnological production processes published in literature (see **Table 2.1** in **chapter 2.1.2**). Therefore, the relevance of further investigations of the production process and possibilities for increased production becomes apparent.

2.1.2 Production of rebeccamycin

Rebeccamycin can be produced in three different ways: Chemical synthesis, heterologous production in other microorganisms or production in the natural producer *L. aerocolonigenes* (Kaneko et al. 1985; Gallant et al. 1993; Faul et al. 1999; Hyun et al. 2003; Casini et al. 2018; Onaka et al. 2015; Walisko et al. 2017). **Table 2.1** gives an overview of the currently achieved rebeccamycin titers and space-time-yields (STY) under various conditions presented in literature and will be considered in more detail below.

Table 2.1: Overview of rebeccamycin concentrations and space-time-yields (STY) achieved with different process conditions from literature. Selected examples based on highest concentration and space-time-yield are presented.

Rebeccamycin concentration [mg L^{-1}]	Organism	Medium	Cultivation volume [L]	T [°C]	Cultivation time [d]	STY [mg L^{-1} d^{-1}]	Comments	Reference
27.1	*Lechevalieria aerocolonigenes*	H34	0.1	27	7	3.87		Bush et al. (1987)
71.3	*L. aerocolonigenes*	H95	0.1	27	7	10.19		Bush et al. (1987)
160.8	*L. aerocolonigenes*	H96	0.1	27	7	22.97		Bush et al. (1987)
183	*L. aerocolonigenes*	H96	0.1	27	7	26.14		Nettleton, Jr. et al. (1985)
0.3-0.5	*Escherichia coli* (recombinant)	LB/ amp	1	26	2	ca. 0.2		Hyun et al. (2003)
0.7	*E. coli* (recombinant)	LB Miller	0.05	26	2	0.35	Rebeccamycin aglycone	Casini et al. (2018)
ca. 8	*Streptomyces lividans (recombinant)*	A-3M	0.1	30	12	0.67	Pure culture	Onaka et al. (2015)
multiplied	*Streptomyces albus* (recombinant)	R5A	0.4	30	5			Sánchez et al. (2002)
663	*L. aerocolonigenes*	G134	30	27	8.5	78.00	Bioreactor scale	Bush et al. (1987)
52.5	*L. aerocolonigenes*	G134			9	5.83		Onaka et al. (2003a)
18.4	*L. aerocolonigenes*	MM + starch	0.1	28	6	3.07		Lam et al. (1989)
44.5	*S. lividans (recombinant)*	A-3M	0.1	30	12	3.71	Co-culture with *T. pulmonis*	Onaka et al. (2015)
ca. 30	*S. lividans (recombinant)*	A-3M	0.1	30	12	2.50	Co-culture with *R. erythropolis*	Onaka et al. (2015)
ca. 15	*S. lividans (recombinant)*	A-3M	0.1	30	12	1.25	Co-culture with *C. glutamicum*	Onaka et al. (2015)
9.6	*L. aerocolonigenes*	GYM	0.05	28	8	1.20	No talc supplement	Walisko et al. (2017)
ca. 18	*L. aerocolonigenes*	GYM	0.05	28	8	2.25	Talc 10 g L^{-1}	Walisko et al. (2017)
94.3	*L. aerocolonigenes*	GYM	0.05	28	8	11.79	Talc $(PDA/Carr)_3$ 10 g L^{-1}	Walisko et al. (2017)
6	*L. aerocolonigenes*	GYM	0.05	28	8	0.75	No glass beads	Walisko et al. (2017)
116	*L. aerocolonigenes*	GYM	0.05	28	8	14.50	Glass beads 80 g L^{-1}, Ø = 0.5 mm	Walisko et al. (2017)

Chemical syntheses

Different chemical syntheses for the rebeccamycin production can be found in literature (Kaneko et al. 1985; Gallant et al. 1993; Faul et al. 1999). The first two approaches of a chemical rebeccamycin production were presented by Kaneko et al. (1985). One approach was a Diels-Alder reaction between N-benzyloxymethylmaleimide and 7,7′-dichloroindigo. The other approach was based on the indole Grignard method, which was later used for further developments. Gallant et al. (1993) improved this total synthesis by a facilitated glycosylation. Faul et al. (1999) introduced a versatile method that allowed the production of either rebeccamycin or its analogues i.e., 11-dechlororebeccamycin. The overall yield for rebeccamycin in this approach was 12 %. Further approaches for a total chemical synthesis of rebeccamycin were scarce, rather a total synthesis for derivatives (Marminon et al. 2003) or chemical transformations of rebeccamycin (Anizon et al. 2009) were presented. A more detailed overview of the chemical syntheses presented in literature is given in Pommerehne et al. (2019).

Heterologous production

The heterologous production of rebeccamycin has various advantages over the natural production in *L. aerocolonigenes*. Challenges by the complex morphology (see also **chapter 2.2**) and the relatively unknown microorganism regarding metabolic pathways and cultivation conditions can be circumvented by the transfer of the rebeccamycin gene cluster into a well-studied model microorganism. Hyun et al. (2003) transferred the gene cluster into *Escherichia coli* and successfully produced rebeccamycin, however, only in small concentrations between 0.3 - 0.5 mg L^{-1} (see **Table 2.1**). Casini et al. (2018) were able to produce the precursor called rebeccamycin aglycon in *E. coli*. A concentration of 0.7 mg L^{-1} was estimated, assuming a similar UV absorbance-to-titer conversion factor for both rebeccamycin and rebeccamycin aglycon. The transformation of a cosmid with the rebeccamycin gene cluster into *Stretompyces lividans* was conducted by Onaka et al. (2003b). A successful rebeccamycin production was confirmed by HPLC measurements but was not quantified. Later on, Onaka et al. (2015) quantified the rebeccamycin production in *S. lividans*. After 12 days of cultivation approximately 8 mg L^{-1} of rebeccamycin were produced. Sánchez et al. (2002) transformed genomic DNA of *L. aerocolonigenes* into *Streptomyces albus* and stated that the production levels were several fold higher than in *L. aerocolonigenes* under the same conditions. However, no numbers were given in this particular case.

Natural production in L. aerocolonigenes

The natural production of rebeccamycin in the original microorganism *L. aerocolonigenes* is described several times in literature. Cultivations after the first isolation resulted in rebeccamycin concentrations between 27.1 and 183 mg L^{-1} in shaking flasks by a variation of the media components (Bush et al. 1987; Nettleton, Jr. et al. 1985). With the H96 medium a rebeccamycin concentration of 160.8 mg L^{-1} after 7 days was achieved in shake flasks. An improved medium with 15 g L^{-1} additional starch (G134 medium) led to 663 mg L^{-1} rebeccamycin after 8.5 days of cultivation in a 30 L bioreactor (Bush et al. 1987), which until now is the highest ever achieved rebeccamycin concentration according to literature (see **Table 2.1**). However, later Onaka et al. (2003a) were not as successful using the same G134 medium. In shake flasks only 52.5 mg L^{-1} were achieved. The different rebeccamycin titers resulting from various media compositions and cultivation parameters already indicate that cultivations of filamentous microorganisms, and thereby their productivity, can be influenced by various parameters. Further insight into this topic is given in **chapter 2.2**.

2.1.3 Purification of rebeccamycin

Rebeccamycin produced during cultivations of *L. aerocolonigenes* or other microorganisms needs to be purified in order to use it for medical purposes. Therefore, the first step is usually the extraction of rebeccamycin from the cultivation broth using tetrahydrofuran (Pearce et al. 1988), acetone (Lam et al. 1989) or ethyl acetate (Walisko et al. 2017). Walisko (2017) added the fatty alcohol ethoxylate Genapol C 200 at the end of the cultivation to facilitate the extraction by avoiding organic solvents. An aqueous extraction of hydrophobic substances like rebeccamycin was enabled using this substance. In literature the description of further purification steps is scarce. Mostly quantitative HPLC analyses are conducted to measure product titers (Walisko et al. 2017; Onaka et al. 2015). However, an HPLC based purification of rebeccamycin with an C18 column and 0.1 M ammonium acetate-MeOH-MeCN (4:3:3) was presented by Pearce et al. (1988). A UV detector was used to detect rebeccamycin at 313 nm. Unfortunately, no further details on the purification process were given. Another purification method is the crystallization of rebeccamycin described by Nettleton, Jr. et al. (1985). Several steps of extraction were done and the final crystallization was conducted with hot methanol.

2.2 Morphology and productivity of filamentous microorganisms

Cell morphology of filamentous microorganisms is more complex than that of unicellular microorganisms. The formation of hyphae that are intertwining to loose mycelia or dense pellets leads to challenges in the cultivation process (Walisko et al. 2015; Wucherpfennig et al. 2010) as is exemplarily shown in **Figure 2.2**. However, these challenges are accepted since many filamentous bacteria and fungi produce substances that are interesting for food or pharmaceutical industry (Walisko et al. 2015). The bacterial filamentous *Actinomycetes*, including *L. aerocolonigenes*, are especially interesting because of the production of mainly pharmaceutically relevant secondary metabolites (Nielsen 1996; Kieser et al. 2000). In the year 2000 60 % of the known biologically active secondary metabolites were isolated from *Actinomycetes* (Kieser et al. 2000). Although there are clear differences between filamentous bacteria and fungi, their morphology and its influence during a cultivation process are similar (Olmos et al. 2013; Nielsen 1996).

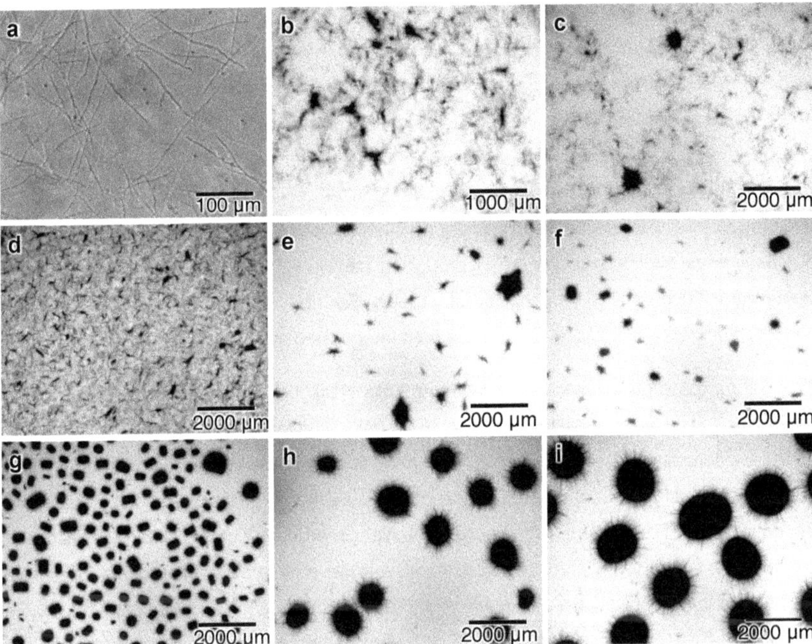

Figure 2.2: Morphologic diversity of the filamentous actinomycete *Lentzea aerocolonigenes* in different cultivations ranging from (a) single hyphae, over mostly mycelial structures (b – d) up to dense and defined pellets in various sizes (e – i).

The different morphological forms of filamentous microorganisms can affect the cultivation process by influencing different parameters such as the viscosity of the cultivation broth and thus the mixing performance as well as the costs for downstream processing (Wucherpfennig et al. 2010). At high biomass concentrations mycelial structures or pellets with many freely exposed hyphae in their periphery can lead to a high viscosity with a non-Newtonian flow behavior (Wucherpfennig et al. 2010; Bliatsiou et al. 2020). In a highly viscous cultivation broth, the mass transfer and mixing performance are impeded. In order to ensure a sufficient nutrient and oxygen supply of the microorganism an increased power input is necessary (Oncu et al. 2007). A large power input in turn causes high mechanical stress and therefore leads to morphological changes (El-Enshasy et al. 2006) as well as variations in biomass growth and product titers (Rosa et al. 2005). Furthermore, dense pellets with a large diameter often face oxygen and nutrient limitations in the pellet core (Hille et al. 2005; Bizukojc and Gonciarz 2015). The diffusion of oxygen in the pellet core is impeded when a critical diameter is exceeded (Hille et al. 2009). This also affects the microorganisms metabolism and the product formation (Phillips 1966; Elmayergi et al. 1973).

For many filamentous microorganisms a correlation between morphology and productivity has been observed (Antecka et al. 2016a). Which morphology is beneficial in regard to productivity, however, highly depends on the microorganism and the desired product. Mycelial structures are advantageous for the production of geldamycin in *S. hygroscopicus* and tylosin in *S. fridae* (Dobson et al. 2008; Tamura et al. 1997). Other microorganisms like *S. tendae* and *S. avermitilis* show an improved production of nikkomycin and avermectin, respectively, in pellets (Vecht-Lifshitz et al. 1992; Yin et al. 2008). However, there are also microorganisms that exhibit nearly equal productivity in different morphological forms like *S. clavuligerus* and *S. virginiae* when producing clavulanic acid or virginiamycin (Belmar-Beiny and Thomas 1991; Yang et al. 1996). For the rebeccamycin production with *L. aerocolonigenes* Walisko et al. (2017) stated the pelleted morphology to be beneficial.

The cell morphology of filamentous microorganisms can be influenced by various process conditions (Oncu et al. 2007; Papagianni 2004; Wucherpfennig et al. 2010; Walisko et al. 2015). By adjusting the cell morphology according to the preferences of the microorganism a targeted optimization of the product concentration is desired. Thus, different aspects need to be considered. For several actinomycetes a high energy dissipation and high dissolved oxygen levels lead to pelleted growth whereas a lower energy dissipation generates fluffier pellets and low dissolved oxygen levels promote mycelial growth (Vecht-Lifshitz et al. 1990; Tough and Prosser 1996; Bellgardt 1998; Yin et al. 2008). Moreover the inoculum size and type, the pH value, the composition of the cultivation medium or the presence of certain ions were observed to influence the morphology and productivity of various filamentous fungi and bacteria (Walisko et al. 2015). The addition of inorganic salts, the so called *salt-enhanced cultivation*, has already been investigated by Wucherpfennig et al. (2011) for cultivations of *A. niger*. The supplementation of sodium chloride led to an increased productivity and a decreased pellet size. Similar observations were made for the addition of ammonium sulfate to *Actinomadura namibiensis* (Tesche et al. 2019). Furthermore, the addition of various other additives to the cultivation can influence the cell morphology. The addition of a silicone antifoam, Tween 80 and Triton X-100 to cultivations of *Streptomyces hygroscopicus* var. *geldanus* resulted in decreasing pellet diameters with increasing concentrations of the corresponding surfactant (Dobson et al. 2008). O'Cleirigh et al. (2005) added xanthan gum to the same microorganism leading to decreased pellet diameters. Another approach, which is already widely applied in fungal cultivations, is the addition of microparticles with a size of up to approximately 50 µm (Antecka et al. 2016a; Karahalil et al. 2019). Walisko et al. (2017) used this method in cultivations of *L. aerocolonigenes* and furthermore applied larger glass particles in a size range of 0.5 - 2 mm. Hence, these aspects will be considered in more detail in the following chapters.

2.2.1 Microparticle enhanced cultivation (MPEC)

One approach to influence the morphology and productivity of filamentous microorganisms is the addition of microparticles, which is meanwhile a frequently used method by researchers (Walisko et al. 2012; Antecka et al. 2016a; Karahalil et al. 2019). This approach of *morphology engineering* is called *microparticle enhanced cultivation* (MPEC) (Kaup et al. 2008). The microparticles applied in these cases usually have a diameter of < 50 µm (Driouch et al. 2011; Gonciarz and Bizukojc 2014; Walisko et al. 2017), in some cases also even larger particles are handled as microparticles (Etschmann et al. 2015; Niu et al. 2020).

The first application of microparticle addition to filamentous microorganisms was described by Kaup et al. (2008). They investigated the effects of aluminum oxide (Al_2O_3) and talc powder ($3MgO \cdot 4SiO_2 \cdot H_2O$) on growth and chloroperoxidase (CPO) formation of *Caldariomyces fumago*. Both types of microparticles had diameters ≤ 42 µm. The microparticle concentration was varied between 0.05 and 25 g L^{-1}. For an aluminum oxide concentration of 15 g L^{-1} a 4-fold increase in accumulated CPO activity was observed after 12 days of cultivation. For 10 g L^{-1} of talc powder even a 10-fold increase of accumulated CPO activity was found. Moreover, the cell morphology of the fungus was significantly influenced by the microparticle addition. Without particles *C. fumago* grew in pellets with a diameter of approximately 4 mm. With aluminum oxide or talc powder the diameter was between 0.1 and 0.5 mm and many single hyphae between 60 and 600 µm were apparent. A similar effect on morphology was also observed for eight other filamentous fungi and the filamentous bacterium *Streptomyces aurefaciens* (Kaup et al. 2008).

Driouch et al. (2010a, 2011) also investigated the effects of talc (ø = 6 µm) and aluminum oxide (ø = 14 µm), but in cultivations of recombinant *Aspergillus niger*. They observed decreasing pellet diameters with increasing microparticle concentration, even leading to freely dispersed mycelium. However, higher concentrations of aluminum oxide were necessary to achieve the same effect as talc at lower concentrations (Driouch et al. 2011). The addition of talc to frucofuranosidase or glucoamylase producing strains of *A. niger* led to a doubled or 4-fold increased enzyme activity, respectively (Driouch et al. 2010a). Since then, many researchers worked with the procedure of MPEC. An overview of different studies applying MPEC is given in **Table 2.2**.

Table 2.2: Overview of selected applications of MPEC in cultivations with different microorganisms from literature.

Microorganism	Microparticle type	Particle conc. [g L^{-1}]	Reference
Caldariomyces fumago	Aluminumoxide, talc	0.05 – 25	Kaup et al. (2008)
Aspergillus niger	Aluminum oxide, talc, titanate	0 – 30	Driouch et al. (2011), Driouch et al. (2012)
Aspergillus niger, Trichoderma atroviride	17 particle types	20	Etschmann et al. (2015)
Streptomyces sp. M-Z18	Talc	0 – 20	Ren et al. (2015)
Aspergillus terreus	Talc	1 – 15	Gonciarz and Bizukojc (2014), Gonciarz et al. (2016)
Aspergillus sojae	Aluminumoxide, talc	0 – 25	Yatmaz et al. (2016), Germec et al. (2017), Karahalil et al. (2017)
Aspergillus niger	Talc	12	Kowalska et al. (2017)
Lentzea aerocolonigenes	Talc, surface modified talc	10	Walisko et al. (2017)
Trichoderma viride	Aluminum oxide	0 – 30	Dong et al. (2018)
Grifola frondosa	Talc	0 – 20	Tao et al. (2018)
Aspergillus terreus	Aluminum oxide	6 – 12	Boruta and Bizukojc (2019)
Aspergillus terreus, Penicillium rubens, Chaetomium globosum, Mucor racemosus	Aluminum oxide	6	Kowalska et al. (2018), Kowalska et al. (2019), Kowalska et al. (2020)
Aspergillus terreus	Talc	0.48	Saberi et al. (2020)
Aspergillus nidulans	Talc	20	Niu et al. (2020)

As can be seen in **Table 2.2** most approaches of MPEC are conducted with filamentous fungi, whereas filamentous bacteria are underrepresented in this field of research. In the here presented literature, the effect of MPEC on filamentous bacteria was only described shortly by Kaup et al. (2008) and in more detail by Ren et al. (2015) and Walisko et al. (2017). Furthermore, talc and aluminum oxide microparticles are applied in nearly all studies (compare **Table 2.2**). Further microparticles were investigated by Etschmann et al. (2015). The effects of 17 microparticle types, differing in size (ø = 5 - 250 µm) and chemical composition, were examined in cultivations of *A. niger* and *Trichoderma atroviride*. However, the effects were highly strain and particle specific and no general mechanistical description was possible (Etschmann et al. 2015).

One of the microparticles used by Etschmann et al. (2015) is titanate, which has already been investigated in detail for cultivations of *A. niger* (Driouch et al. 2012). Titanate ($TiSiO_4$, ø = 8 µm) microparticles led to a different effect on morphology than previously described. The particles were incorporated into the pellets resulting in a core of microparticles at larger particle concentrations. The overall size of the pellets remained similar at low particle concentrations up to 5 g L^{-1} and was decreasing for larger concentrations. However, no freely dispersed mycelium was generated. Despite the different effects on morphology, the fructofuranosidase activity was increased 4-fold under titanate addition (Driouch et al. 2012).

Etschmann et al. (2015) proposed the design of tailor-made particles with a defined surface, since the inhomogeneity regarding particle size and shape of different microparticle types might play a major role regarding their effects on cultivations of filamentous microorganisms. Walisko et al. (2017) applied a layer-by-layer surface modification technique on talc particles. Thus, talc particles with different surface modifications and thereby different surface properties were created. These surface modified talc particles were then used in cultivations of *L. aerocolonigenes*. 10 different modifications together with unmodified talc were investigated. The production of rebeccamycin with some modified particles was higher than with unmodified talc, in other cases the product concentrations were lower. Rather hydrophilic or only slightly hydrophobic particles with a negative zeta potential showed the highest rebeccamycin concentrations. However, no clear trend of which properties are advantageous could be identified (Walisko et al. 2017).

The mechanisms behind MPEC are widely discussed, but not yet fully understood. Driouch et al. (2010a) traced the effects of talc particles on morphology of *A. niger* back to the hindered spore aggregation before germination. The different effects of talc or aluminum oxide were explained by the different size and material of both microparticles as a major factor. Electron microscopic images showed rather smooth and round-shaped aluminum particles while talc particles were irregular and inhomogeneous (Driouch et al. 2011). The proposed reason for the increase in enzyme activity with talc or aluminum oxide was the loosened mycelial structure, that allows an increased nutrient and oxygen supply to all areas (Driouch et al. 2010a). Although titanate addition led to another morphology, still a loosening of the pellet structure was observed, leading to similar improvements in nutrient and oxygen supply throughout the pellet (Driouch et al. 2012). Talc and titanate microparticles were also added to a cultivation of the recombinant reporter strain *A. niger* ANip7-MCS-gfp2 producing a variant of green fluorescent protein (GFP) which was coexpressed with the desired enzyme glucoamylase. Larger areas of active biomass contributing to enzyme production were observed in case of talc and titanate addition. (Driouch et al. 2010b; Driouch et al. 2010a; Driouch et al. 2012). While a pellet from an unsupplemented cultivation was only producing in a thin layer on the surface of the pellet, pellets grown with titanate produced GFP all over the pellet (Driouch et al. 2012). Mycelial structures resulting from talc addition also showed production in all areas (Driouch et al. 2010b).

The idea of hindered spore aggregation was further investigated by Kowalska et al. (2020). They investigated the effect of aluminum oxide on cell morphology in four different fungi *A. terreus, P. rubens, C. globosum and M. racemosus* with different mechanisms of morphology formation: spore agglomerative, hyphae agglomerative, spore and hyphae agglomerative and non-agglomerative, respectively. The largest effects on morphology were observed with the purest agglomeration types, spore agglomerative *A. terreus* and non-agglomerative *M. racemosus*. For *C. globosum* a new morphological star shape of pellets was observed, while the effects on *P. rubens* were comparably low. The results indicate, the mechanism of agglomerate formation plays a major role with microparticle addition (Kowalska et al. 2020).

2.2.2 Macroparticle-enhanced cultivation

Since the addition of microparticles proved to be beneficial for many filamentous microorganisms, some authors questioned whether larger particles might have similar effects. Kaup et al. (2008) tested quarz (SiO_2) particles with a diameter of approximately 350 µm and glass beads with a diameter of about 500 µm in cultivations of *C. fumago*. The quarz particles showed a similar effect as the microparticles for biomass and CPO formation, but much less pronounced. The glass beads, however, did not affect biomass and CPO formation in the cultivation (Kaup et al. 2008). The addition of broken and porous SiO_2 particles (ø = 120 – 200 µm) and glass beads (ø = 250 – 500 µm) to two different *Streptomyces* strains was investigated by Holtmann et al. (2017). 5 g L^{-1} of SiO_2 in a cultivation of *Streptomyces coelicolor* led to an increase in actinorhodin concentration of 85 % up to 160 % depending on the applied medium. The final streptavidin concentration after a cultivation of *Streptomyces avidinii* was not affected by the addition of 5 g L^{-1} glass beads. However, an accelerated production was observed (Holtmann et al. 2017). Ochi (1986) used 100 of 3 mm glass beads for the homogenization of spores before inoculation and also applied glass beads for the homogenization of biomass.

Neither of the aforementioned authors investigated the morphological changes caused by the addition of larger particles. Ochi (1986) have described the homogenization of biomass, but have not given any further details on pellet sizes. Hotop et al. (1993) added 4 mm glass beads to their pre-cultures (50 in the first pre-cultures, 400 in the second pre-cultures) to reduce the pellet diameter of *Penicillium chrysogenum*. The pellet diameter in the pre-cultures was reduced to < 300 µm, leading to a diameter reduction from 1 to 0.6 mm in the main culture. The final product concentration doubled (Hotop et al. 1993). Glass bead addition was furthermore investigated by Dobson et al. (2008). The glass beads with a diameter of 5 mm were used in a 250 mL shaking flask for the cultivation of *Streptomyces hygroscopicus var. geldanus*. Between 0 and 55 glass beads were added at the beginning and a spore suspension was used for inoculation. With increasing glass bead number, the average pellet diameter decreased. These changes in cell morphology were accompanied by an increased final geldamycin concentration with increasing glass bead number (Dobson et al. 2008). Sohoni et al. (2012) applied glass beads between 0.75 and 4 mm in microtiter plate cultivations of *S. coelicolor.* Diameters of 0.75 - 2 mm led to pelleted growth, whereas diameters of 3 and 4 mm generated dispersed mycelium. A glass bead diameter of 3 mm was considered optimal, as it led to reproducible and narrow sized cell morphology of *S. coelicolor* and in addition improved the product concentrations of actinorhodin and undecylprodigiosin (Sohoni et al. 2012). The addition of glass beads to *L. aerocolonigenes* was described by Walisko et al. (2017). Glass bead diameters between 0.25 – 2.1 mm in a concentration of 80 g L^{-1} were added to 250 mL baffled shaking flasks. In Walisko (2017) even larger particles up to 5 mm were tested in cultivations. With increasing glass bead diameter, the pellets diameter decreased resulting in a mycelial structure with glass beads of ≥ 3 mm. With 0.25 - 0.5 mm glass beads a 19-fold increase in rebeccamycin concentration compared to an unsupplemented control was observed (Walisko 2017; Walisko et al. 2017).

The mechanical stress induced by the glass beads was often considered to be responsible for the effect in cell morphology, since mechanical stress by aeration or agitation can affect pellet morphology as well (Lin et al. 2010). Thus, Walisko et al. (2017) made a rough estimation of the mechanical stress induced by glass beads based on a stress model for grinding in a ball mill. In this model Kwade (2003) determined two characteristic values, the stress energy (equation (1)) and the stress frequency (equation (2)), describing the comminution in stirred media mills.

$$SE \propto SE_{gm} = d_{gm}^3 \cdot u_t^2 \cdot \rho_{gm} \tag{1}$$

The stress energy (SE) quantifies the maximum amount of energy transferred to the product particle in a single stress event. In the equation d_{gm} is the diameter of the grinding media, u_t is the stirrer tip speed and ρ_{gm} is the density of the grinding media.

$$SF = \frac{SN}{t} \tag{2}$$

The stress frequency (SF) describes the number of stress events (SN) per time. The product of stress energy and stress frequency is proportional to the specific energy input (equation (3))

$$E_{spec} \propto SE \cdot SF \tag{3}$$

However, as stated above this approach can only be used for a rough estimation since the conditions in a stirred media mill essentially differ from those in a shake flask. For example, the degree of filling with grinding media differs, which in the case of the stirred media mill is usually 70 - 85 % of the grinding chamber (Kwade and Schwedes 1997), and in the shaking flask cultivations mentioned above only comparatively low quantities of glass beads were applied (Walisko et al. 2017). For a more precise estimation of the mechanical stress induced by macroparticles in shaking flask cultivations, CFD-DEM-simulations of the system similar to the approach of Beinert et al. (2015; 2018) for stirred media mills are necessary. A first approach of CFD-DEM coupled simulations for the stress estimation in cultivations of *L. aerocolonigenes* with glass macroparticles is presented by Schrader et al. (2019).

2.2.3 *In-situ* product recovery with adsorbent resins

Another particle type that is occasionally added to cultivations of many filamentous microorganisms are adsorbent resins. These adsorbent resins can be applied for an *in-situ* product removal (ISPR) since the desired product can be adsorbed by the resins. However, additionally to this effect, an increased product titer was often observed (Phillips et al. 2013). Different mechanism for these increased product concentrations by adsorbent resins were described in literature. On the one hand the adsorption of the desired product to the resin can reduce effects of product degradation (Tsueng and Lam 2007), feedback inhibition (Kim et al. 2011) or autotoxicity during the cultivation (Singh et al. 2010). On the other hand disadvantageous media components or by-products can be adsorbed and an accumulation in the liquid phase is avoided (Yu et al. 2002; Park et al. 2007). However, in some cases the resins also adsorb essential media components which can be a drawback (González-Menéndez et al. 2014).

In 1959 first approaches with the addition of ion exchange resins during the production of neomycin und novobiocin was conducted. A binding of the product was possible; however, an increased productivity was not observed. In the 1970's macroporous non-ionic polymeric adsorbent resins were developed and gained interest for their application in cultivations with microorganisms (Phillips et al. 2013). Many different adsorbent resins with different properties exist and were successfully applied in microbial cultivations. In a cultivation of *Actinoplanes teicomyceticus* the addition of the adsorbent resin Diaion HP-20 and Amberlite XAD-16 led to an adsorption of the product teicoplanin and had a positive effect on the teicoplanin concentration. However, Diaion HP-20 addition resulted in the highest product titer with a 4.2-fold increase compared to an unsupplemented cultivation. The elimination of toxic effects and feedback inhibition of teicoplanin was assumed to be the reason for this increased production (Lee et al. 2003). The addition of 20 g L^{-1} Amberlite XAD-4, Amberlite XAD-7 or Amberlite XAD-16 to a cultivation of *Salinospora tropica* all resulted in an increased titer of the product NPI-0052. The largest enhancement was observed with Amberlite XAD-7 addition, which resulted in a 69-fold increase. In this case Amberlite XAD-7 stabilized the product NPI-0052 by adsorption (Tsueng and Lam 2007). Cultivations with other filamentous microorganisms such as *Penicillium chrysogenum* or *Streptomyces platnesis* also resulted in increased product titers when adsorbent particles were added (Warr et al. 1996; Woo et al. 2002).

Phillips et al. (2013) reviewed this topic and listed almost 20 different particle types with their properties. Various adsorbent particle types made of different kinds of material, with different particle sizes, surface areas and pore sizes are listed. Furthermore, three aspects to consider for adsorbent resin addition to a cultivation are mentioned: the type of adsorbent resin, the resin concentration and the time of addition to the culture (Phillips et al. 2013).

A further aspect to consider was studied by Frykman et al. (2006). If the adsorbent resins are added to a cultivation in a shaking flask or a bioreactor, they are constantly exposed to mechanical stress. Breakage of the adsorbent resins can occur leading to an impeded recovery of the beads at the end of a cultivation, since the resins are often separated from the cultivation broth with a sieving mesh. Amongst others, Frykman et al. (2006) compared bead-bead and bead-agitator collisions in a bioreactor as possible means of resin breakage. Mathematical models were used to simulate the collisions. These models indicated, that the resin breakage is more likely to be caused by bead-agitator collisions rather than bead-bead collisions. An experimental evaluation using XAD-16 resin in a water-glycerol solution was performed and particle size distributions were measured using laser diffraction. After 3 days at a stirrer speed of 600 min^{-1} (impeller tip speed = 2.0 m s^{-1}) no significant changes in resin size were observed. Thereafter, the impeller speed was increased to approximately 800 min^{-1} (impeller tip speed = 2.7 m s^{-1}). Subsequently, a decreasing particle size was determined, decreasing further the longer the stirring process was conducted (Frykman et al. 2006). Cultivations for antibiotic production in larger scale are usually performed with impeller tip speeds between 5 and 7 m s^{-1} (Junker 2004). Hence, this topic needs to be considered when applying adsorbent resins in microbial cultivations.

2.2.4 Cell morphological characterization and pellet viability

Since the cell morphology is an important parameter for cultivations of filamentous microorganisms its investigation is a crucial analytical aspect. The most applied method for the characterization of cell morphology, according to Papagianni (2014), is the acquisition of light microscopic images followed by (semi) automated-image analysis (e.g. Willemse et al. (2018) and Cairns et al. (2019)). Depending on the desired parameters of micro- or macro-morphology different methods are applied (Krull et al. 2013). Conventional microscopic images only allow the analysis of the outer macro-morphology, however, pellet slicing with subsequent microscopy additionally enables the characterization of the inner pellet structure (Lin et al. 2010). The disadvantage of this method is the high manual effort of pellet slicing leading to a low number of pellets for analysis and therefore to a low statistical validity

(Schrinner et al. 2020). For small mycelial structures, even a micro-morphological characterization is possible with parameters like the total hyphal length, number of hyphal tips, hyphal growth unit or number of branching points (Barry et al. 2015). A relatively new approach for the characterization of the inner pellet structure and the micro-morphology inside pellets is based on X-ray microcomputed tomography. However, until now this method is only applicable for microorganisms with a hyphal diameter > 3 µm (Schmideder et al. 2019a).

The micro- and macro-morphology of pellets should not be considered separately, but together with the impact of the structure on e.g. substrate transport or pellet viability which are important for an overarching understanding. If the micro-morphology inside the pellet is characterized, the substrate transport inside the pellet can be modelled. This has already been conducted for pellets of *A. niger* (Schmideder et al. 2019b). The viability of pellets or different pellet areas can be obtained by viability staining in combination with fluorescence microscopy (Nieminen et al. 2013; Rajnisz et al. 2016). Rajnisz et al. (2016) even used 5-cyano-2,3-bis(4-methylphenyl)-2H-tetrazolium chloride (CTC) to make the respiration activity of *Streptomyces* sp. 8812 visible with a fluorescence microscope. The viability obtained by fluorescent staining in these cases is determined *off-line*. However, flow cytometry enables the quantitative determination of pellet viability and cell morphology is additionally considered in an *at-line* process. With low measurement times larger amounts of pellets are measured resulting in a high statistical robustness. Moreover, flow cytometry measurements have the potential to be performed in an *on-line* process in the future (Ehgartner et al. 2016; Ehgartner et al. 2017; Veiter and Herwig 2019).

2.3 Bubble–free aeration in bioreactors

In conventional stirred tank bioreactor cultivations of bacteria or fungi the aeration is conducted with different types of gas spargers releasing bubbles. These bubbles are then dispersed in the cultivation medium by the stirrer (Belmar-Beiny and Thomas 1991; Tamura et al. 1997; Lin et al. 2010; Garcia-Ochoa et al. 2013). Despite the simplicity of this technique there are several disadvantages. The aeration with bubbles induces stress in the cultivation broth e.g., by bubble rupture at the surface of the liquid (Henzler 2000). Lin et al. (2010) investigated the effects of aeration and agitation induced volumetric power input during cultivations of the fungus *A. niger*. The fraction of aeration or agitation induced volumetric power input was varied, while the total volumetric power input was kept constant. An increased share in the aeration induced volumetric power input led to smaller pellets and a higher pellet concentration whereas contrary results were found for an increased share in the agitation induced volumetric power input. Thus, aeration induced a higher mechanical stress than agitation and thereby influenced growth and cell morphology of *A. niger* to a greater extent (Lin et al. 2010). In summary, the mechanical stress induced by aeration is not negligible and should be considered especially for cultivations of filamentous microorganisms. This is the reason why bubble free aeration systems are often applied for cultivations of mammalian cells which are sensitive towards mechanical stress (Frahm et al. 2009). Furthermore, bubbles often generate foam during the cultivation. Next to the aeration, media components, cell growth and formation of metabolites or surface-active substances can promote foaming (Taticek et al. 1991; Vardar-Sukan 1992). Extensive foaming can be a problem in cultivations since a loss of biomass by flotation effects, substrate or product from the cultivation broth can occur (Junker 2007). The addition of antifoam agents decreases the gas-liquid volumetric mass transfer coefficient (Morão et al. 1999) and may inhibit growth or production (Berovič and Cimerman 1979). Thus, the use of a bubble-free aeration system seems advantageous for stress-sensitive cultivation systems.

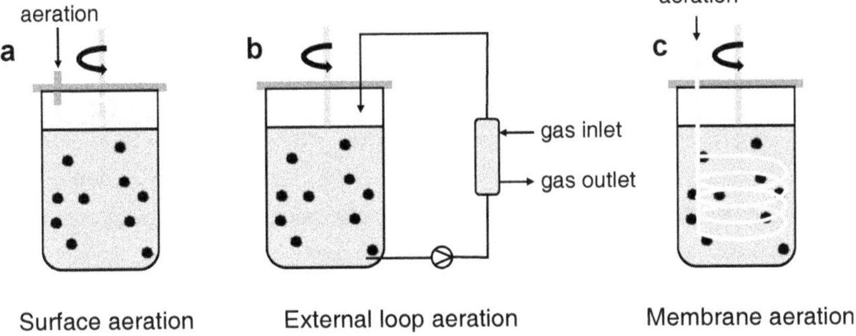

Figure 2.3: Three different types of bubble-free aeration techniques in bioreactor scale: (a) surface aeration, (b) external loop aeration and (c) membrane aeration.

In **Figure 2.3** three different techniques for bubble-free aeration are schematically displayed; surface aeration, aeration through an external loop and membrane aeration. The surface aeration is a very commonly used type of aeration. However, it is mainly applied in shake flask cultivations (Yang and Liau 1998; Maier and Büchs 2001) or microtiter plates (Doig et al. 2005). With larger cultivation volumes the surface aeration alone is not sufficient for an adequate oxygen supply. The relation of liquid height and liquid surface is increased leading to a decreased specific transfer surface (Kopp 1989). Another bubble-free aeration technique is the use of an external loop. In this case the medium from the bioreactor is pumped into an external aeration module, where it is enriched with oxygen, and is then pumped back into bioreactor (Kopp 1989; Schneider et al. 1995). There are, however, some drawbacks of this method that need to be considered. The medium should be pumped through the loop without cells to avoid cell growth in the aeration module (Kopp 1989). For the use with filamentous microorganisms the pumping of cells could furthermore influence the growth or cell morphology since mechanical stress is induced (Kamilakis and Allen 1995). Moreover, the oxygen input in the entire cultivation volume depends on the volume flow rate of the pump. An efficient applicability of such a system was therefore not considered to be practical (Kopp 1989). A further possibility for bubble-free aeration is membrane aeration. For this method an oxygen permeable membrane, often in form of tubes, is introduced in the cultivation medium. Air or oxygen is then led through this permeable membrane and is thereby introduced in the cultivation broth (Lehmann et al. 1988; Schneider et al. 1995; Qi et al. 2003). This low-shear, bubble-free cultivation technique has often been successfully applied in cultivations of mammalian cells (e.g. Luttmann et al. (1994) and Schneider et al. (1995)). Since the membrane aeration showed general applicability in cultivations, it is discussed in more detail in the following chapter.

2.3.1 Membrane aerated bioreactors

Membrane aeration offers many parameters that can be varied to achieve the desired goal. First of all, the membrane material has to be chosen carefully. In general, two different types of material are distinguished: A hydrophobic microporous membrane, such as polypropylene (PP) (Lehmann et al. 1988) or Polytetrafluoroethylene (PTFE) (Schneider et al. 1995), as well as a non-porous, but oxygen permeable membrane, such as silicone (Luttmann et al. 1994). Due to the hydrophobicity of the microporous membranes no cell growth on the membrane appears which is an important advantage over silicone. Although the membrane is porous, usually no cultivation medium will enter through the pores. This is only possible if its pressure exceeds the liquid entry pressure. However, some gasses, such as oxygen, can diffuse through the pores of the membrane into the cultivation broth. The driving force for this process is the partial pressure difference between both sides of the membrane (Wagner and Lehmann 1988; Schneider et al. 1995). Schneider et al. (1995) compared the mass transfer from silicone based membrane aeration from literature with their results using PTFE tubing and found the microporous PTFE tubing to be more effective. However, the geometry in the different reactor systems was different. Hence, this comparison gives only an idea of the differences between these two membrane materials (Schneider et al. 1995). Nevertheless, the hydrophobic microporous membranes have a disadvantage over the non-porous silicone membranes. If the pressure on the inside of the membrane is too high (above the bubble-point), bubbles can be detached from the pore surface and the aeration is no longer bubble-free (Wagner and Lehmann 1988).

The way of installation of the membrane and the membrane dimensions are also an aspect of interest. Schneider et al. (1995) presented different reactor configurations with PTFE membranes. In one case a flat membrane was integrated at the base of the bioreactor and used for aeration. In another configuration a flat sheet membrane was used in an external loop. A third approach used a PTFE tube inserted into the liquid (Schneider et al. 1995). The application of membrane tubes, however, is most commonly described in literature (Lehmann et al. 1988; Luttmann et al. 1994; Qi et al. 2003; Frahm et al. 2009). The tubular membrane can be installed in form of a single long tube or several shorter tubes. Installing only one very long tube leads to differences in pressure along the length of the tube. At the gas inlet the pressure is highest and it is lowest at the tube exit (Qi et al. 2003). Hence, in some cases the tube installed in the bioreactor is divided in several shorter segments to reduce this effect (Frahm et al. 2009). The tubing applied is mostly thin-walled and preferably long to achieve a large surface exchange area (Wagner and Lehmann 1988; Lehmann et al. 1988; Luttmann et al. 1994; Schneider et al. 1995; Qi et al. 2003).

Another aspect to consider is whether the membrane is installed in a static manner or dynamically. In earlier years of membrane aeration, the membranes were mostly installed with a fixed position in the bioreactor and were not moved during a cultivation (Wagner and Lehmann 1988; Luttmann et al. 1994; Schneider et al. 1995; Qi et al. 2003). Lehmann et al. (1988) and Frahm et al. (2009) described a dynamic membrane aeration by additionally using the installed membrane as a stirrer. In this way aeration and stirring are combined. Frahm et al. (2009) described a crucial effect of this method when a stirrer composed of silicone tubes was installed in a bioreactor: Through the movement of the silicone tubes no cell growth on the membrane appeared during cultivation. In this case the silicone tubes were wound in order to form kind of large stirrer blades. This stirrer was then oscillated to move 180 ° in one direction and 180 ° back. This dynamic membrane aeration was investigated in reactors of different volumes. The volumetric mass transfer coefficient of these systems was clearly increased compared to a conventional static membrane aeration (Frahm et al. 2009).

2.3.2 Oxygen transfer in bioreactors

In aerobic cultivation processes, one of the most important substrates is oxygen which is necessary for cell growth, cell maintenance and product formation. However, due to the low solubility of oxygen in aqueous solutions a sufficient supply of the microorganisms can be challenging. Hence, oxygen needs to be continuously introduced in the cultivation broth (Garcia-Ochoa et al. 2010).

To achieve a sufficient oxygen supply for microbial growth, the estimation of the oxygen transfer rate (OTR) under different process conditions and in different scales is important. Knowledge of the OTR is relevant for the selection, design and scale-up of bioreactors. The OTR is proportional to the concentration difference of oxygen at the gas-liquid boundary layer and the bulk phase as well as to the volumetric mass transfer coefficient (k_La) which can be determined from experimental measurements by different methods. These values are essential to assess the aeration efficiency in bioreactors (Garcia-Ochoa and Gomez 2009).

The oxygen level in the liquid phase of a bioreactor results from the continuous oxygen input by aeration of the culture medium (oxygen transfer rate, OTR) and the oxygen consumption or uptake (oxygen uptake rate, OUR) by the microorganisms that are growing in the culture medium. This relation can be described with equation (4).

$$\frac{dc_L}{dt} = OTR - OUR \tag{4}$$

In case of the absence of microorganisms during the experimental measurement no oxygen consumption occurs (OUR = 0). This can be used for a cell-free characterization of the aeration efficiency in a bioreactor. The simplest and most used theory for the description of the gas-liquid mass transfer is the two film model. Taking this into account, equation (4) can be simplified to equation (5):

$$\frac{dc_L}{dt} = OTR = k_L a \cdot (c_L^* - c_L) \tag{5}$$

with k_La being the volumetric mass transfer coefficient, c^*_L being the oxygen saturation concentration in the liquid phase and c_L the oxygen concentration in the liquid phase. When the oxygen concentration in the liquid phase (c_L) is 0, the oxygen transfer rate reaches its maximal value (OTR_{max}) as given by equation (6).

$$OTR_{max} = k_L a \cdot c_L^* \tag{6}$$

Since a sufficient oxygen input for a non-substrate limiting cultivation is one of the greatest challenges for bioprocess technology, cultivations are also carried out with oxygen-enriched air as well as under positive pressure to increase the oxygen solubility (Qi et al. 2003; Frahm et al. 2009).

3 Material and methods

3.1 Strain and cultivation medium

Lentzea aerocolonigenes DSM 44217 was used for the investigations in this thesis. It was purchased from the German Collection of Microorganisms and Cell Cultures (DSMZ, Braunschweig, Germany).

The cultivation medium was GYM (Glucose-Yeast-Malt) medium consisting of 4 g L^{-1} glucose, 4 g L^{-1} yeast extract and 10 g L^{-1} malt extract. Before autoclaving the pH was set to 7.2 using 2 M KOH.

3.2 Particles

3.2.1 Surface modified microparticles

Surface modified microparticles were prepared by M. Schrader (Institute for Particle Technology (iPAT), TU Braunschweig, Germany). Modifications were done on micro glass beads (Mirco Glass Beads, Sigmund Lindner GmbH, Warmensteinach, Germany). The microparticle size range given by the manufacturer and a median particle size are given in **Table 3.1**. For simplification, the microparticles will be referred to by the given median particle size herein after.

Table 3.1: Particle size range as given by the manufacturer and median particle size x_{50} of the micro glass beads used for modifications.

Particle size range [µm]	Median particle size x_{50} [µm]
0-20	7.9
0-50	30.5

The modification generally described by e.g. Kockmann et al. (2016) was modified for this purpose and was conducted in three steps. First, the micro glass beads were pretreated with NaOH. 50 g of microparticles were suspended in 200 mL 4M NaOH and heated at 100 °C for 15 min in an oil bath. Subsequently the suspension was centrifuged for 5 min at 8000 min^{-1} (Avanti JXN-26, Beckman Coulter, Brea, USA), the supernatant was decanted and the microparticles were washed with water three times. Then the particles were dried at 120 °C overnight. The second step was the silanization of the microparticles using (3-aminopropyl)triethoxysilane (APTES). The reaction was conducted in ethanol with a total volume of 20 mL. 0.5 g mL^{-1} microparticles were suspended in approximately two-thirds of the ethanol, while 0.6 mL_{APTES} $g^{-1}_{particle}$ were dissolved in the remaining ethanol. The APTES-solution was added to the microparticle suspension and the mixture was heated to 100 °C in an oil bath for 20 h under constant mixing and using a reflux condenser. Subsequently, the suspension was centrifuged for 10 min at 8000 min^{-1}, the supernatant was decanted and the microparticles were washed with ethanol three times. The microparticles were then dried at room temperature overnight. The third reaction step was the functionalization using selected carboxylic acids (see **Table 3.2**). The amount of carboxylic acid used in this reaction, was chosen to correspond the amount of substance used during the silanization ($2.56 \cdot 10^{-3}$ mol_{APTES} $g^{-1}_{particle}$). 0.1 g mL^{-1} microparticles were used in this reaction step. The carboxylic acid was first dissolved in one-third of the final solvent volume. Diisopropylcarbodiimid (DIC) was added in a molar ratio of 1.5 as activator. The microparticles were suspended in the remaining solvent and subsequently mixed with the activated carboxylic acid. This mixture was stirred for 20 h at room temperature. Afterwards centrifugation and washing were conducted as already described for the silanization with APTES. Microparticles were finally dried overnight at room temperature before their use in cultivations.

Table 3.2: Carboxylic acids with their corresponding molecular formula and chemical structure as used for surface modifications of glass microparticles.

Carboxylic acid	Abbreviation	Molecular formula	Chemical structure
Butanoic acid (Butyric acid)	BA	$C_4H_8O_2$	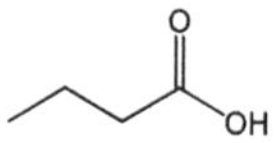
Hexanoic acid (Caproic acid)	HA	$C_6H_{12}O_2$	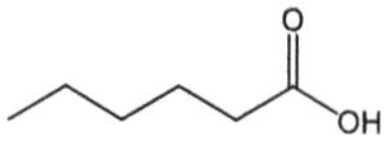
Decanoic acid (Capric acid)	DA	$C_{10}H_{20}O_2$	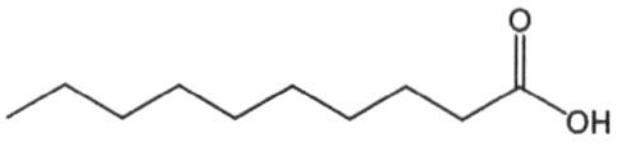
3-Mercaptopropionic acid	MPA	$C_3H_6O_2S$	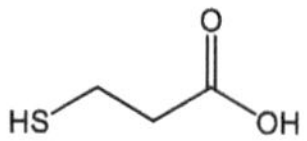
3-Phenylpropanoic acid (Hydrocinnamic acid)	PPA	$C_9H_{10}O_2$	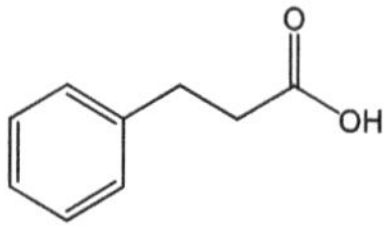
Hexanedioic acid (Adipic acid)	HDA	$C_6H_{10}O_4$	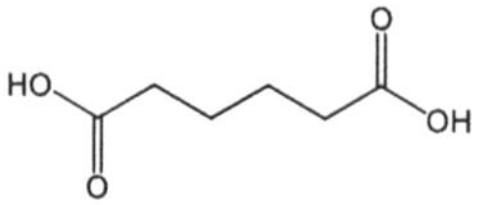

3.2.2 Macroparticles

Different macroparticles of various size ranges were used in the cultivations. All macroparticles were purchased from the Sigmund Lindner GmbH (Warmensteinach, Germany). Glass beads (type S and micro glass beads) and ceramic beads (type Z) were used. Further details on material and size are found in **Table 3.3**. For simplification, the macroparticles will be referred to by the given mean particle size herein after.

Table 3.3: Particle material, size range as given by manufacturer and mean particle size $\bar{x}_{3,2}$ (see Schrader et al. (2019) for calculation) of supplemented particles.

Material	Particle size range [µm]	Mean particle size $\bar{x}_{3,2}$ [µm]
Glass beads (ρ = 2500 kg m^{-3})	200-300	282
	400-600	540
	500-750	658
	750-1000	969
	1000-1300	1183
	1250-1650	1513
	1550-1850	1746
	1700-2100	1932
Ceramic beads (ρ = 3800 kg m^{-3})	800-1000	918

3.2.3 Adsorbent particles for *in-situ* product removal

Particles with adsorbent properties were applied in shake flask cultivations. The effects of three different kinds of resins Amberlite ® XAD 4, Amberlite ® XAD 7 HP and Amberlite ® XAD 16 N (all purchased from Sigma-Aldrich, Steinheim, Germany) during the cultivation were investigated.

Before being used in cultivations, the resins were washed with water by keeping it in Milli-Q (Gradient A10, Millipore, Eschborn, Germany) overnight. The water was removed the next morning. This step was performed to remove salts in which the resins are embedded to avoid bacterial growth during storage.

The particle sizes were measured using laser diffraction (Mastersizer 2000, Malvern Panalytical GmbH, Kassel, Germany) for verification of the manufacturers information. The particles were suspended in distilled water in a glass beaker on a magnetic stirrer plate. The suspension was pumped (Ismatec ® ISM444B, Cole-Parmer GmbH, Wertheim, Germany) through the measuring cell and the amount of particles was adjusted to fit an obscuration of

5 - 30 %. The results are mean values (n = 5) of a volume-based distribution. The particle suspensions were measured again after 24 h of stirring in the glass beaker.

3.2.4 Adsorption of rebeccamycin to adsorbent particles

The culture broth of an 8 day cultivation was treated with an ultra-turrax (T25, IKA Labortechnik, Staufen, Germany) at 13500 min^{-1} for 4 min. The biomass was then separated from the cultivation broth. The particles were prepared as described above and 50 g L^{-1} of each particle type were added to 50 mL of the rebeccamycin containing supernatant in a shake flask. Additionally, an unsupplemented control was prepared. The shake flask was incubated for three days at 120 min^{-1} and 28 °C in a rotary shaker (Certomat BS-1, Sartorius, Göttingen, Germany) to allow adsorption under cultivation conditions (see **chapter 3.3**). Subsequently, the particles were separated from the supernatant by centrifugation for 10 min at 4000 min^{-1} (Heraeus Varifuge 3.0R, Thermo Fisher Scientific, Waltham, USA). Rebeccamycin was extracted separately from the supernatant and the particles as described in **chapter 3.4.3**.

3.3 Cultivation methods

3.3.1 Cryopreservation and inoculation

For the cryoculture preparation of *Lentzea aerocolonigenes* 50 mL of GYM medium were filled in a 250 mL baffled shake flask and inoculated with 1 mL cell suspension from a cryoculture. This shake flask was incubated at 28 °C and 120 min^{-1} (Certomat BS-1, Sartorius, Göttingen, Germany) for 2 days in darkness. 500 µL of the cell suspension were mixed with 500 µL 60 % (v/v) glycerol and frozen at -80 °C.

The inoculation of main cultures requires a pre-culture which is performed in a similar way as described above. The baffled shake flask with 50 mL of GYM medium is inoculated with the 1 mL cryoculture and incubated for 2 days at 28 °C and 120 min^{-1}. A defined volume depending on the cultivation volume of this pre-culture is then used for the inoculation of the main culture.

3.3.2 Shaking flask cultivations

Cultivation in shake flasks were performed in 250 mL shake flasks with four baffles. The filling volume was 50 mL (GYM medium). Main cultures were inoculated with 300 µL of a pre-culture (see **chapter 3.3.1**) and incubated at 28 °C and 120 min^{-1} (Certomat BS-1, Sartorius, Göttingen, Germany) for 10 days in darkness, if not stated otherwise. The orbital shaker amplitude was 50 mm.

For the screening of many different growth conditions, samples were only taken and analyzed at the end of the cultivation (after 10 days if not stated otherwise). These experiments were always performed in triplicates. For the data acquisition of growth kinetics complete shake flasks were removed from the shaker and used for sampling every one to two days in order not to change the filling volume and therefore oxygen transfer by sampling. Since larger amounts of shake flasks are necessary in this case, those experiments were conducted in duplicates.

When particles were added, they were only used in the main culture. Macroparticles were weighed into the shaking flasks and autoclaved. Afterwards, 50 mL medium (see **chapter 3.1**) were added and the cultivation was inoculated and incubated as described above. The surface modified microparticles or resins were also weighed into the shaking flasks but were suspended in 10 mL purified water before autoclaving. After autoclaving 40 mL concentrated medium (see **chapter 3.1**) were added to the shaking flasks (the medium was prepared with 20 % less water). Inoculation and incubation were performed as described above. The weight of the supplemented particles is indicated with the corresponding results.

Supplementation with soy lecithin (MP Biomedicals, Eschwege, Germany) was prepared similarly. A stock solution of 50 g L^{-1} was prepared and mixed with water to set the desired concentration in the shaking flask. 40 mL of a concentrated GYM medium were then added to result in a total filling volume of 50 mL. Inoculation and incubation were performed as described above.

For the pre-incubation of the cultivation medium with adsorbent particles, the shake flasks were prepared as described above. The flasks were shaken for 2 days at cultivation conditions (28 °C, 120 min^{-1}, 50 mm amplitude). After this time the content of the shaking flasks were transferred to 50 mL tubes and centrifuged for 15 min at 4000 min^{-1} (Heraeus Varifuge 3.0R, Thermo Fisher Scientific, Waltham, USA). Additionally, the supernatant was filtered with a syringe tip filter (Minisart RC 0.2, Sartorius, Göttingen, Germany) to remove remaining particles. The particle-free medium was transferred in shake flasks, inoculated and incubated as described above.

3.3.3 Bioreactor cultivations

Cultivations in larger scale were carried out in 3 L stirred tank bioreactors (Applikon, Schiedam, The Netherlands) with a working volume of 1.2 L (GYM medium, see chapter **3.1**). An elephant ear impeller (down-pumping; EED) was used. Bioreactors were inoculated with 7.5 mL of a pre-culture (see chapter **3.3.1**) and incubated at 28 °C for 10 days. Aeration and agitation speed were varied in different experiments and are indicated with the corresponding results. Aeration was conducted with pure oxygen. Bioreactor cultivations were conducted in duplicates if not stated otherwise.

A silicone-based membrane aeration system was developed during this thesis to create a bubble-free aeration inside the bioreactor. The silicone tubes listed in **Table 3.4** were used for aeration in the bioreactor.

Table 3.4: Dimensions and manufacturer of the installed silicone tubes for membrane aeration (d_i = inner diameter, d_o = outer diameter, s = wall thickness, A_m = membrane exchange surface).

d_i [mm]	d_o [mm]	s [mm]	A_m [cm^2]	Manufacturer
2.5	2.9	0.2	196 (2.5 m tube) 393 (5 m tube)	RCT Reichelt Chemietechnik GmbH & Co.KG, Heidelberg, Germany
4.6	5.2	0.3	max. 451 (filling volume dependent)	Haarmann GmbH & Co. KG, Mittenaar, Germany

Two different membrane aeration systems were used in this thesis. The first membrane aeration system consisted of a silicone tube with 2.9 mm outer diameter, which was installed in the bioreactor by winding it around the reactor installations. The tube was installed to generate three modules of similar sizes with free spaces in between (see **Figure 3.1**). The length of the tube was varied and was either 2.5 or 5 m long. The applied length is indicated with the corresponding results. The open end of the tube was directed in the headspace to create an additional headspace aeration.

Figure 3.1: (a) Installation of the thin-walled silicone tube (d_o = 2.9 mm, d_i = 2.5 mm; s = 0.2 mm) in the bioreactor. Modules vary in size depending on tube length (b) 5 m and (c) 2.5 m.

The second membrane aeration system was designed and constructed in cooperation with M. Schrader and R. Rösemeier-Scheumann (iPAT, TU Braunschweig, Germany) to guarantee a defined geometry (**Figure 3.2**). In this case larger silicone tubes (d_o = 5.2 mm, d_i = 4.6 mm) were used for aeration. 26 silicone tubes were installed vertically and fixed with hose clamps.

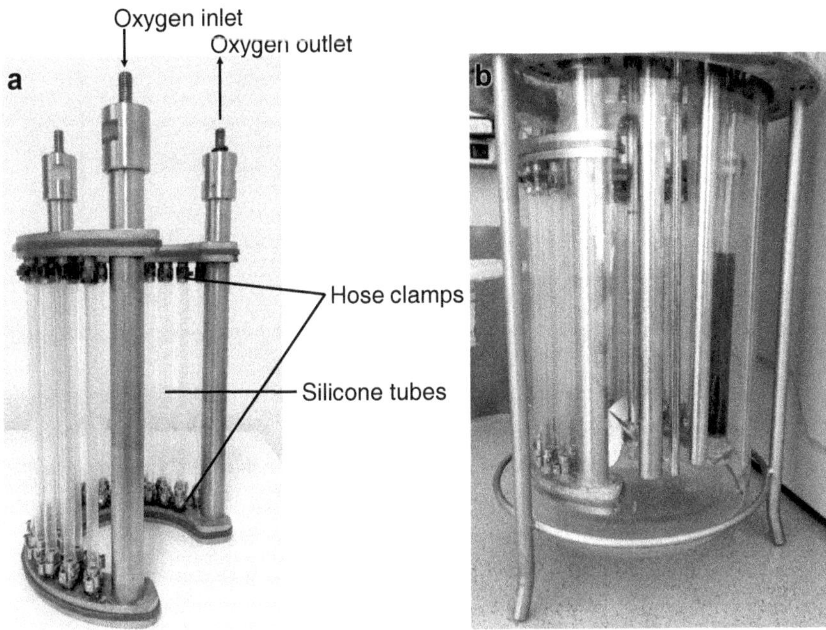

Figure 3.2: Membrane aeration system with 26 vertically installed thin-walled silicone tubes (d_o = 5.2 mm, d_i = 4.6 mm), (a) structure of the membrane aeration system and (b) installation in the bioreactor.

For aeration oxygen is introduced through the inlet and is led into the lower C-shaped component through a pipe. From there the oxygen is distributed into the silicone tubes, where it can diffuse into the cultivation medium. Excess oxygen enters the upper C-shaped component and can be released through the outlet. The construction can be installed in the bioreactor and the shape leaves enough space for further installations like O_2- and pH-electrodes and a sampling tube.

During cultivations pressure was applied through the installation of a valve behind the outlet. A pressure gauge was additionally installed to be able to set a specific pressure value (**Figure 3.3**). The hose clamps shown in **Figure 3.2** were added to keep the silicone tubes in place under pressure.

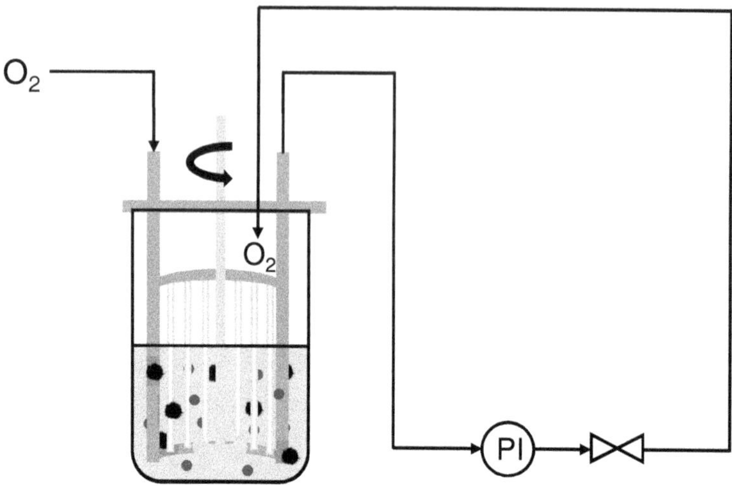

Figure 3.3: Schematic representation of the oxygen flow in the reactor with a pressure gauge and valve to apply pressure to the aeration system.

The oxygen flow through the membrane aeration system with an additional pressure gauge and valve is depicted in **Figure 3.3**. Oxygen is inserted in the inlet of the membrane aeration system and excess oxygen, which is not diffusing through the membrane, is released through the outlet. It is then led to the pressure gauge and valve through an oxygen impermeable tube. The valve is adjusted manually and the pressure gauge shows the corresponding pressure. The adjusted pressure is indicated with the corresponding results. Any oxygen released by the valve is returned to the headspace of the bioreactor to generate an additional headspace aeration.

3.4 Analytical methods

3.4.1 Cell dry weight concentration determination

Cell dry weight concentration (CDW) was determined gravimetrically. Paper filters (Quantitative Papers/Grade 389, Sartorius, Göttingen, Germany) were dried at 105 °C for 24 h, subsequently cooled down in a desiccator and weighed on a precision scale (CP225D, Sartorius, Göttingen, Germany). A defined volume of the sample was filtered, dried at 105 °C for 48 h, cooled down and weighed again.

Large macroparticles sedimented quickly resulting in no particles being removed during sampling. Since small microparticles show slow sedimentation, another procedure for the separation of biomass and microparticles was developed. A sieve with an outer diameter of 50 mm, a mesh size of 50 µm and a frame of 1 cm height was built. This sieve was placed into a petri dish, a defined volume of the culture was transferred onto the sieve and deionized water was added. This set-up was shaken (Titramax 1000, Heidolph Instruments, Schwabach, Germany) for 15 min. The biomass stays in the sieve while the free microparticles accumulate in the petri dish. Both were collected separately in ceramic cups and dried at 105 °C for 24 h. To learn about the amount of particles that were incorporated in the pellets an additional incineration was conducted. The dried biomass in the ceramic cups was incinerated at 550 °C for 2 hours in a muffle furnace (Heraus M110, Fisher Scientific GmbH, Schwerte, Germany). After opening the muffle furnace to place the ceramic cups inside, additional 10 min are necessary to reach the correct temperature of 550 °C.

3.4.2 Glucose quantification

The cultivation broth was filtered (0.2 µm pore size, PTFE membrane syringe filter, VWR, Darmstadt, Germany) to obtain a cell-free supernatant. The glucose concentration in the supernatant was determined via HPLC (HitachiLaChrom Elite, Hitachi, Tokyo, Japan). The eluent was highly purified water (Advantage A10, Millipore, Eschborn, Germany) used at a flow rate of 0.6 mL min^{-1}. For separation the pre-column Metacarb 87C Guard (50 x 4.6 mm, AgilentTechnologies, Santa Clara, USA) and the column Metacarb 87C (300 x 7.8 mm, Agilent Technologies, Santa Clara, USA) were utilized at 85 °C. The glucose was detected by a reflective index detector (Hitachi L-2490 RI detector, Hitachi, Tokyo, Japan). The retention time with this HPLC method for glucose was about 10.5 min.

3.4.3 Rebeccamycin extraction and quantification

Rebeccamycin was extracted from the cultivation broth with ethyl acetate. 5 mL ethyl acetate were added to 10 or 20 mL (bioreactor or shake flask scale, respectively) of cultivation broth and the sample was then incubated in an overhead shaker (Intelli-Mixer RM-2 M, LTF Labortechnik, Wasserburg, Germany) for 60 min in darkness. Subsequently, the sample was centrifuged at 4000 min^{-1} for 10 min (Heraeus Varifuge 3.0R, Thermo Fisher Scientific, Waltham, USA) to enhance phase separation and the organic phase was removed for HPLC measurements (HitachiLaChrom Elite, Hitachi, Tokyo, Japan). The Hypersil ODS pre-column (5 µm, 50 × 4.6 mm, Thermo Fisher Scientific, Waltham, USA) and the Hypersil ODS column (5 µm, 250 × 4.6 mm, Thermo Fisher Scientific, Waltham, USA) were utilized for separation at 30 °C with a flow rate of 1.0 mL min^{-1}. The eluents 0.1 % trifluoric acid (TFA) and 90 % acetonitrile (ACN) were applied according to the gradient described in **Table 3.5**. Rebeccamycin was detected with a diode array detector (Hitachi L-2455 Diode Array Detector, Hitachi, Tokyo, Japan) at a wavelength of 316 nm. Rebeccamycin was detected after a retention time of approximately 18 min.

Table 3.5: Eluent gradient for the rebeccamycin quantification by HPLC.

Time [min]	0.1 % trifluoric acid [%]	90 % acetonitrile [%]
0	77.8	22.2
20	16.6	83.4
22	16.6	83.4
23	77.8	22.2
30	77.8	22.2

3.4.4 Amino acid quantification

The samples for amino acid quantification were filtered (0.2 µm pore size, PTFE membrane syringe filter, VWR, Darmstadt, Germany) to obtain cell-free supernatant. The amino acid concentration was determined via HPLC measurements (Agilent Series 1200, Agilent Technology, Waldbronn, Germany). A pre-column (Gemini "C18", MAX-RP, 4 mm x 3 mm, Phenomenex, Aschaffenburg, Germany) was used for derivatization with o-phthaldialdehyde and a reversed phase Gemini column (5 µm C18 110A, 150 x 4.6 mm, Phenomenex, Aschaffenburg, Germany) was used for the separation of amino acids. Measurements were conducted at 40 °C. Two eluents A (40 mM NaH_2PO_4, pH 7.8 adjusted with 5 M NaOH) and B (methanol : acetonitrile : water; 45 : 45 : 10) with an overall flow rate of 1 L min^{-1} were applied according to the gradient described in **Table 3.6**. A fluorescence detector (340 nm excitation, 540 nm emission; Agilent, Waldbronn, Germany) was used for detection. Tryptophan was detected after a retention time of approximately 36 min.

Table 3.6: Gradient of eluent A (40 mM NaH_2PO_4, pH 7.8 adjusted with 5 M NaOH) and eluent B (methanol : acetonitrile : water; 45 : 45 : 10) for the quantification of amino acids by HPLC.

Time [min]	Eluent A [%]	Eluent B [%]
0.0	100.0	0.0
40.5	59.5	40.5
41.0	39.0	61.0
43.0	39.0	61.0
57.5	0.0	100.0
59.5	0.0	100.0
60.5	25.0	75.0
61.5	50.0	50.0
62.5	75.0	25.0
63.5	100.0	0.0
65.5	100.0	0.0

3.4.5 Determination of the volumetric mass transfer coefficient (k_La)

The volumetric mass transfer coefficient (k_La) was determined using the dynamic gassing-out method for shaking flasks and the stirred tank bioreactor. In both cases the corresponding volume of GYM medium was filled in and used for cell-free measurements.

For recording changes in the dissolved oxygen (DO) in shaking flasks a shake flask reader (PreSens Precision Sensing GmbH, Regensburg, Germany) was utilized. The medium in the shake flask was aerated with nitrogen for approximately 5 min and subsequently, the nitrogen in the headspace was exchanged by air. The shake flask was then incubated in an orbital shaker on the shake flask reader according to the cultivation conditions until the dissolved oxygen reached its maximum.

The DO in the stirred tank bioreactor was recorded using the ArcAir Software and a corresponding mobile device (Hamilton Germany GmbH, Germany) which were collecting data every 3 s. The bioreactor was aerated with nitrogen until the DO was close to 0 % and then aerated with air or pure oxygen until 100 % of saturation was reached. Stirrer speed and the type of aeration were varied for different measurements which are indicated with the corresponding results.

The k_La is mathematically defined according to equation (7) (Garcia-Ochoa and Gomez 2009).

$$\frac{dDO(t)}{dt} = k_La \cdot (DO_{max} - DO(t)) \tag{7}$$

Integration and conversion of the equation leads to the following relationship:

$$ln(DO_{max} - DO(t)) = -k_La \cdot t + C \tag{8}$$

By plotting ln (DO_{max} – DO(t)) against the time, a straight line is created whose slope equals the k_La. All measurements used for the calculation of the k_La were conducted in triplicates.

3.4.6 Calculation of maximal oxygen transfer rate

The calculation of the maximal oxygen transfer rate (OTR_{max}) was conducted according to equation (6). The k_La necessary for the calculation was determined with the dynamic gassing-out method as described in **chapter 3.4.5**. The oxygen saturation concentration was estimated depending on the medium composition according to Wilhelm et al. (1977), Weisenberger and Schumpe (1996), and Rischbieter et al. (1996).

3.5 Investigation of the pellet structure

3.5.1 Microscopy and image analysis

Microscopic images of the biomass were taken with a digital inverse microscope (EVOS XL, AMG, Bothell, WA, USA). The sample was diluted in a petri dish (94 × 16 mm, polystyrene, Greiner Bio-One, Kremsmünster, Austria) and a sufficient number of images were acquired to guarantee at least 500 pellets for characterization, mostly even more. Image analysis was done using the public domain software Fiji/ImageJ 1.52h (National Institute of Health, Bethesda, USA). The microscopic images were converted into 8-bit grayscale images. Background subtraction followed by an enhancement of the contrast were applied for a facilitated separation of the pellets from the background. In the next step, the images were binarized using the automatic thresholding. The pellet morphology was analyzed with the built-in particle analysis function which delivers parameters such as projected area, perimeter, Feret-diameter, aspect ratio, circularity, roundness and solidity. Pellets with a projected area < 1000 µm^2 and those intersecting the edge of an image were excluded from analysis.

In one case, the image analysis was conducted in Matlab (version 2019a, MathWorks, Natick, MA) as part of the cooperation with S. Schmideder (Chair of Process Systems, School of Life Sciences, TU München) (see **chapter 4.2.3**). The microscopic images were binarized with *adaptthresh* in which the threshold is chosen based in the local mean intensity in the neighborhood of each pixel. Pellets lying close together were segmented by watershed segmentation. The Matlab function *regionprops* was used for the determination of morphological parameters such as projected area, perimeter, circularity and Feret-diameter. Small objects, defined to have a projected area < 200 µm^2, and pellets intersecting the edge of an image were deleted automatically.

The morphological parameters Area-Equivalent-Spherical-Diameter (AESD) and circularity are defined as followed:

$$AESD = \sqrt{\frac{4 \cdot Area}{\pi}} \tag{9}$$

$$Circularity = 4\pi \cdot \frac{Area}{Perimeter^2} \tag{10}$$

The AESD is a parameter to describe the pellet size. The circularity describes the shape of pellets. The values of circularity range between 0 and 1, where larger values correspond to rounder pellets. In some cases, another parameter, the roundness (equation (11)), can be considered. Both parameters describe similar properties of the pellets, however, they are mathematically differently defined.

$$Roundness = 4 \cdot \frac{Area}{\pi \cdot Major\ Axis^2} \quad (11)$$

3.5.2 Pellet slicing

For the investigation of the inner pellet structure, a pellet slicing technique was applied. For sample preparation single pellets were frozen in a frozen section medium (Richard-Allan Scientific Neg-50 Frozen Section Medium, Thermo Fisher Scientific, Waltham, USA) on top of a sample holder. Freezing was shortened by using a rapid freezing station in a cryostat microtom (HM 550, Microm, Neuss, Germany). Slices of 100 µm thickness were cut with a knife installed in the cryochamber. Pellet slices were then placed on slides and images were recorded using a microscope (Evos XL, AMG, Bothell, WA, USA).

3.5.3 Viability staining and CLSM

To investigate pellet viability during the cultivation viability staining was applied. The staining solution consisted of 1.5 µL 3.34 mM SYTO9 and 1.5 µL 20 mM propidium iodide (PI) (Molecular Probes, Eugene, USA) in 1 mL of 1x salt solution (**Table 3.7**).

Table 3.7: Composition of the 1x salt solution used for viability staining.

Concentration [g L^{-1}]	Components
0.31	Ammonium chloride
4.33	Disodium phosphate
0.13	Potassium chloride
3.04	Sodium dihydrogen phosphate dihydrate

Single pellets were transferred to the staining solution and incubated for 15 min in darkness. After incubation pellet slicing according to **chapter 3.5.2** was performed. The pellet slices were then investigated using a confocal laser scanning microscope (CLSM, C2si, Nikon Instruments, Amsterdam, Netherlands). The living parts of the pellets were observed with a 510 - 540 nm laser and showed green fluorescence due to staining with SYTO9. Dead parts of the pellets, appearing red by the propidium iodide staining, were observed with a 620 - 650 nm laser. Both stains intercalate with nucleic acid but only SYTO9 can permeate cell membranes whereas PI is impermeable. Therefore, SYTO9 stains all cells containing nucleic acid. PI has a higher affinity towards nucleic acid replacing SYTO9 in dead cells when both stains are present (Stocks 2004). The software Fiji/ImageJ 1.52h (National Institute of Health, Bethesda, USA) was used for image analysis. The live ratio of pellets was calculated using the number of pixels of living (green) and dead (red) parts of the pellets. Therefore, the plugin voxel counter was applied. The pellet live ratio was calculated according to equation (12).

$$Live\ ratio = \frac{Green\ pixels}{Green\ pixels + Red\ pixels} \tag{12}$$

3.5.4 Pellet viability via flow cytometry

Flow cytometric measurements were conducted by L. Veiter (Institute of Chemical, Environmental and Bioscience Engineering, Research Area Biochemical Engineering, TU Wien) (see also Schrinner et al. (2020)). The cultivation sample was first diluted 1:10 in a phosphate buffered salt solution (**Table 3.8**) and stained with 20 µM PI (Sigma Aldrich, St. Louis, Missouri/USA).

Table 3.8: Composition of the phosphate buffered salt solution for dilution in flow cytometry measurements.

Concentration [g L^{-1}]	Components
50	2.65 g L^{-1} $CaCl_2$ solution
0.2	KCl
0.2	KH_2PO_4
0.1	$MgCl \cdot 6\ H_2O$
8	NaCl
0.764	$Na_2HPO_4 \cdot 2\ H_2O$

After 1 min of incubation, further staining with 5 mg L^{-1} fluorescein diacetate (Sigma Aldrich, St. Louis, Missouri, USA) was conducted and incubated for 5 min. For flow cytometric analysis the stained sample was furthermore 1:100 diluted with the same phosphate buffered solution. The green fluorescence is caused by fluorescein diacetate (FDA) by esterase activity indicating metabolic activity. PI causes red fluorescence by intercalating with DNA of cells with damaged membranes (Pekarsky et al. 2018).

Flow cytometry measurements were conducted with a CytoSense flow cytometer CytoBuoy, Woerden, The Netherlands) as described previously (Pekarsky et al. 2018; Veiter and Herwig 2019). Detailed information on the measurement settings in this special case can be found in Schrinner et al. (2020). The software CytoUSB was used for measurements and CytoClus4 (CytoBuoy, Woerden, The Netherlands) was used for data analysis. Signal from four different channels were obtained: sideward scatter (SSC), forward scatter (FSC), green fluorescence (FLG) and red fluorescence (FLR) channel. Further details on the flow cytometry method can be found in Schrinner et al. (2020) and Veiter and Herwig (2019).

Certain morphological parameters of pellets can be determined by flow cytometry. The signal length can be used to describe the pellet diameter. However, the determination of pellet size is limited. Only pellets larger than 80 µm are recognized, smaller objects are not considered to be pellets. Moreover, due to the limited size of the sampling tube, pellets with a diameter larger than 500 µm are excluded and the measurement of pellets with a diameter above 300 µm can be hindered. Furthermore, the compactness, which describes the density of a pellet, can be calculated with the sideward scatter (SSC) signal length and the pellet diameter:

$$Compactness_{SSC} = \frac{Length\ of\ SSC\ signal}{Signal\ length} \tag{13}$$

The length of the SSC signal is given by the signal length at half of the maximum of the signal level of the signal profile. As increased signals appear at large pellet sizes, "Length of SSC signal" is divided by "Signal length" to compensate this effect.

Next to the pellet morphology, physiological aspects can be assessed by flow cytometry. The pellet viability is assessed relying on individual pellet signal shapes. The contribution of the green fluorescence signals (FDA staining) to viability can be obtained by equation (14).

$$Viable\ layer\ vl_{FDA} = \frac{Area\ under\ FLG\ curve}{Area\ under\ FSC\ curve} \cdot signal\ length \cdot 0.5 \tag{14}$$

To consider red fluorescence signals (PI staining), the calculation of the viable layer relies on the individual maximum of fluorescence signals. This maximum is included as a "threshold" value in equation (15) which is defined as 0.3 of the maximum fluorescence.

$$Viable\ layer\ vl_{PI} = 0.5 \cdot (1 - Length\ of\ FLR > threshold) \tag{15}$$

The factor 0.5 is included in equations (14) and (15) to estimate the viable layer as a simplified radius as the pellet is considered to be spherical.

For the estimation of pellet viability, a viability factor was calculated (equation (16)). In this factor the relation of the viable layer (equations (14) and (15)) to pellet size is considered.

$$Viability\ factor = \frac{2 \cdot viable\ layer}{pellet\ size} \tag{16}$$

Furthermore, the autofluorescence of the biomass was determined by measurements without fluorescent staining. The autofluorescence was assessed by calculation of the autofluorescence factor (equation (17)).

$$Autofluorescence\ factor = \frac{Area\ under\ FLG\ curve}{Area\ under\ FSC\ curve} \tag{17}$$

4 Results and discussion

4.1 Microparticle enhanced cultivation

The filamentous actinomycete *Lentzea aerocolonigenes* is a rather unexplored bacterium. As many other filamentous microorganisms, it is able to form hyphae that can enlarge to mycelium or even dense pellets. *L. aerocolonigenes* mainly gained interest because of one of its metabolites rebeccamycin which is a promising antitumor substance (Bush et al. 1987). However, research on the natural production of rebeccamycin, on an understanding of the growth or techniques for process optimization is scarce.

The addition of microparticles to the cultivation of filamentous microorganisms is a widely applied method for morphology engineering (Kaup et al. 2008; Antecka et al. 2016a; Walisko et al. 2015; Karahalil et al. 2019). Walisko et al. (2017) used different microparticles in cultivations of *L. aerocolonigenes* with a significant increase in rebeccamycin titers for 10 g L^{-1} talc particles.

4.1.1 Glass microparticle concentration and growth kinetics

For the following investigations micro glass beads were used as microparticles. Although most microparticles described in literature were talc, aluminum oxide or titanate (Kaup et al. 2008; Driouch et al. 2012; Antecka et al. 2016b; Gonciarz and Bizukojc 2014; Kuhl et al. 2020), these micro glass beads were chosen as they are chemically and thermally very stable. To test for the principle applicability and effects on production of rebeccamycin in *L. aerocolonigenes*, micro glass beads in two different sizes of 7.9 µm and 30.5 µm and different concentrations between 0 and 10 g L^{-1} were investigated in a cultivation. The resulting rebeccamycin concentrations are displayed in **Figure 4.1**.

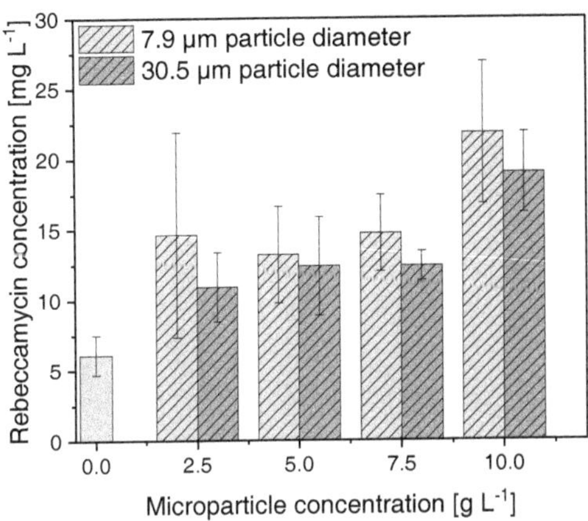

Figure 4.1: Rebeccamycin concentrations of a 10 day cultivation with microparticle addition of x_{50} = 7.9 and 30.5 µm in concentrations between 0 and 10 g L^{-1}.

In all cases of microparticle addition an increased rebeccamycin concentration compared to an unsupplemented control (without particles) can be observed. With increasing microparticle concentration the rebeccamycin concentration increases for both particle sizes. The differences in rebeccamycin titers resulting from the different particle sizes are small, but the mean values for 7.9 µm are always slightly higher than for 30.5 µm particles. At a particle concentration of 10 g L^{-1} the 7.9 µm particles led to around 22 mg L^{-1} of rebeccamycin, whereas 30.5 µm particles resulted in about 19 mg L^{-1} rebeccamycin.

In literature, microparticle addition to cultivations of filamentous microorganisms is often accompanied by changes in the morphological structure (Kaup et al. 2008; Driouch et al. 2011). Hence, the macro-morphology of *L. aerocolonigenes* was investigated and the resulting area-equivalent-spherical-diameters (AESD) of pellets for each approach are presented in **Figure 4.2**. Other morphological parameters like solidity, aspect ratio and roundness did not show significant differences and are therefore not displayed.

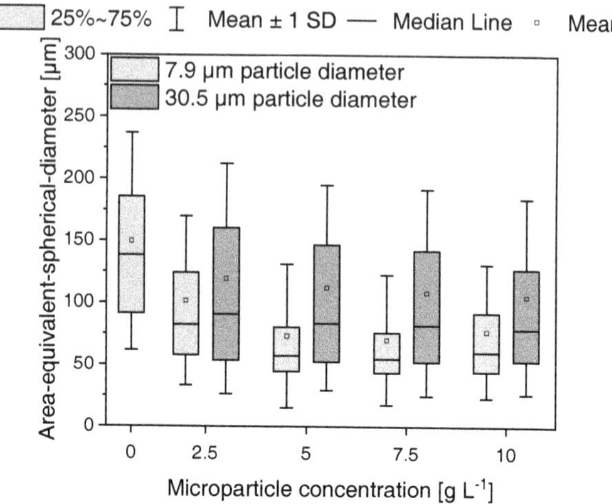

Figure 4.2: Area-equivalent-spherical-diameter (AESD) of pellets grown with microparticle addition (x_{50} = 7.9 and 30.5 µm) in different concentrations of 0 - 10 g L^{-1}.

A lower AESD of pellets from microparticle supplemented cultivations compared to the pellets from an unsupplemented control is visible. However, *L. aerocolonigenes* still appears as more or less distinct pellets, even though very small (see **Figure 4.3**). Moreover, the pellets from cultivations with 30.5 µm particles are slightly larger than those from cultivations with 7.9 µm particles. Furthermore, the tendency of smaller values with increasing microparticle concentration is observed.

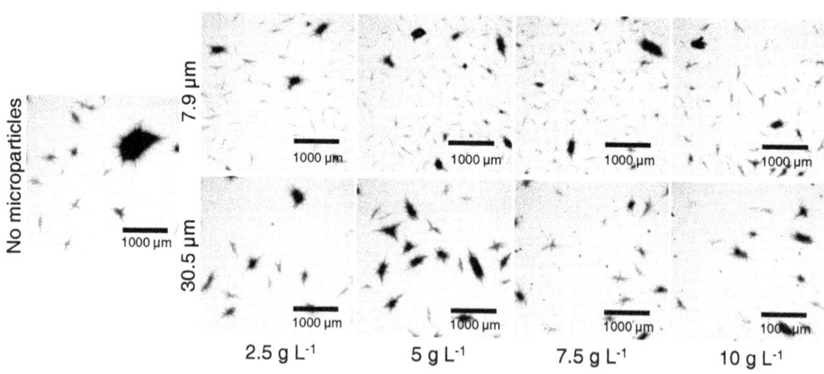

Figure 4.3: Microscopic images of *L. aerocolonigenes* under microparticle addition (x_{50} = 7.9 and 30.5 µm) in different concentrations of 0 - 10 g L^{-1}.

The pellets were not only characterized regarding their macro-morphology, but were additionally sliced to learn about the effects of microparticle addition to the inner structure of these pellets. In **Figure 4.4** pellet slices of every approach are illustrated.

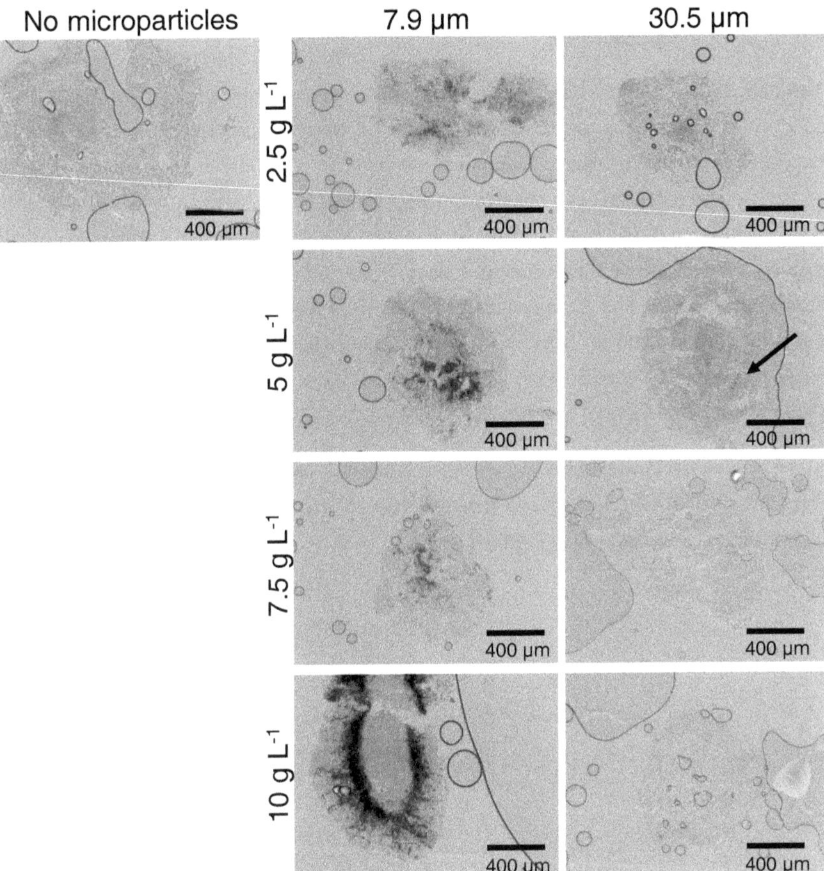

Figure 4.4: Microscopic images of pellet slices from cultivations with micro glass beads (x_{50} = 7.9 and 30.5 µm) in different concentrations from 0 - 10 g L^{-1}. Black lines or circles are air bubbles occasionally enclosed in the sectioning medium. Dark areas in the pellet slices with 7.9 µm glass beads indicate microparticle incorporation. For 30.5 µm glass beads incorporated microparticles are marked with an arrow for improved visibility.

A slice of a pellet grown without microparticle treatment is displayed on the left for comparison. For the addition of 7.9 µm particles the pellet slices clearly show an incorporation of the microparticles into the pellets. The particles are visible as darker areas inside the pellet. This incorporation is visible for all concentrations applied and a tendency of the incorporation of larger numbers of particles in case of larger particle concentrations can be observed. However, for larger particles of 30.5 µm this effect cannot be observed to the same extent. Larger circles in the images as seen for 2.5, 7.5 and 10 g L^{-1} are air bubbles that can be enclosed in the sectioning medium and should not be confused with particles. In this case the only particle incorporation is visible in the image of 5 g L^{-1} (marked with an arrow). Only very few particles are incorporated suggesting that particles are only incorporated until a certain size.

The incorporation of microparticles has already been observed by Walisko et al. (2017). They used 10 g L^{-1} of talc particles in a cultivation of *L. aerocolonigenes* and noticed particles entangled in the pellets exposed hyphae on the outer surface. Furthermore, an incorporation of titanate microparticles was observed in *A. niger* cultivations leading to the formation of core-shell pellets (Driouch et al. 2012). Kuhl et al. (2020) observed the incorporation of talc microparticles in pellets of *Streptomyces albus*. Since 30.5 µm particles were only marginally incorporated in pellets of *L. aerocolonigenes*, the increased rebeccamycin titer might rather be caused by mechanical stress as further discussed in **chapter 4.2** for larger particles. However, since the incorporation of microparticles is an interesting aspect only 7.9 µm particles are used in further cultivations.

In the next step a microparticle (x_{50} = 7.9 µm) supplemented and unsupplemented cultivation were investigated over time with daily sampling to find out about any effect on kinetics. The cultivation results are shown in **Figure 2.1**.

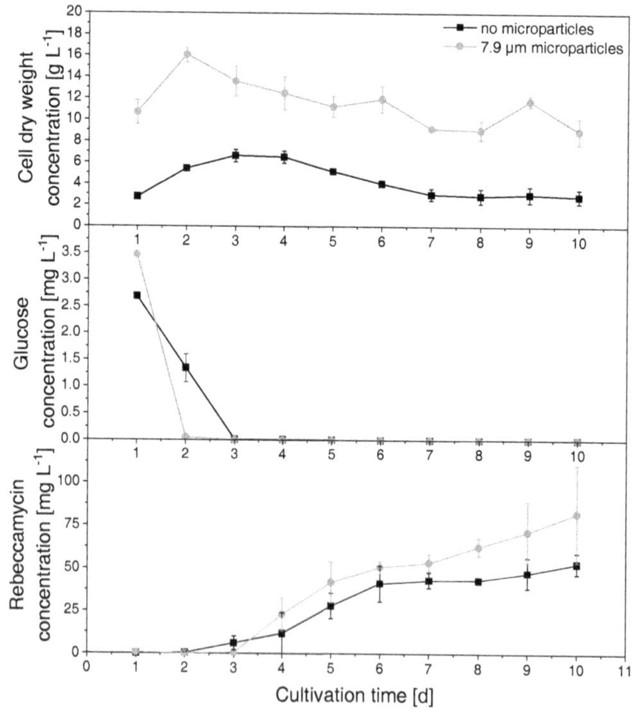

Figure 4.5: Cell dry weight, glucose and rebeccamycin concentrations of a 10 day cultivation without and with 10 g L^{-1} of 7.9 µm glass beads.

When looking at the cell dry weight concentration (CDW), the significantly higher CDW of the microparticle supplemented cultivation is particularly striking. However, this effect is caused by the method for CDW determination. In this case paper filters were used to separate biomass from the cultivation broth. This way not only biomass but also the microparticles were retained. Therefore, unlike the general trend, the total values should not be compared. The course of both CDWs is similar, an increase in the beginning until day 2 or 3 and a decrease thereafter. The glucose is nearly totally consumed after 2 days for the microparticle supplemented cultivation, whereas the glucose in the unsupplemented cultivation is only consumed after 3 days. The rebeccamycin concentration also shows a similar course for both approaches. However, the rebeccamycin formation without microparticles starts between 2 and 3 days while the formation with microparticles starts after 3 days. Despite the decelerated rebeccamycin formation with microparticle addition, the rebeccamycin concentration increases to higher levels of rebeccamycin than in the unsupplemented control. After 10 days of cultivation approximately 53 mg L^{-1} rebeccamycin were produced without microparticles and 82 mg L^{-1} rebeccamycin with microparticle addition.

As already described above, the AESD with microparticle addition was clearly decreased compared to pellets from an unsupplemented cultivation as can be seen in **Figure 4.6**.

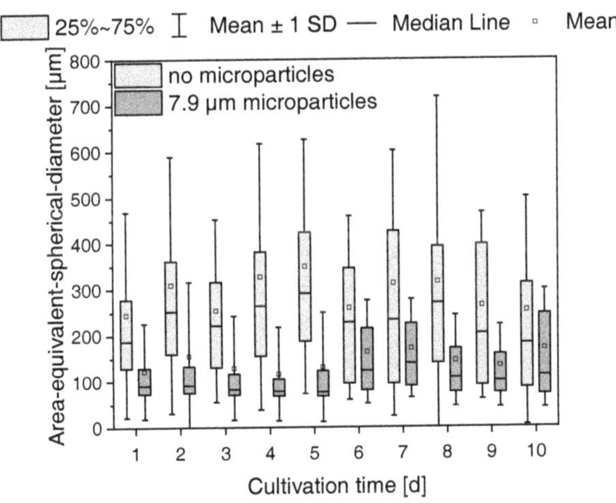

Figure 4.6: Area-equivalent-spherical-diameter of pellets grown without and with microparticle addition (x_{50} = 7.9 µm) during a cultivation of 10 days with daily sampling.

Especially in the first 5 days the differences in the AESD between both approaches are quite large. In addition, a considerably narrower distribution of the AESD in the case of microparticle addition can be seen in this period. The microparticles seem to lead to more homogenous pellet sizes in the cultivation.

To further investigate the effect of microparticles on *L. aerocolonigenes*, viability staining with subsequent pellet slicing was conducted. The share of the living area of a pellet in the total pellet area was calculated and is presented as the live ratio (equation (12)) in **Figure 4.7**.

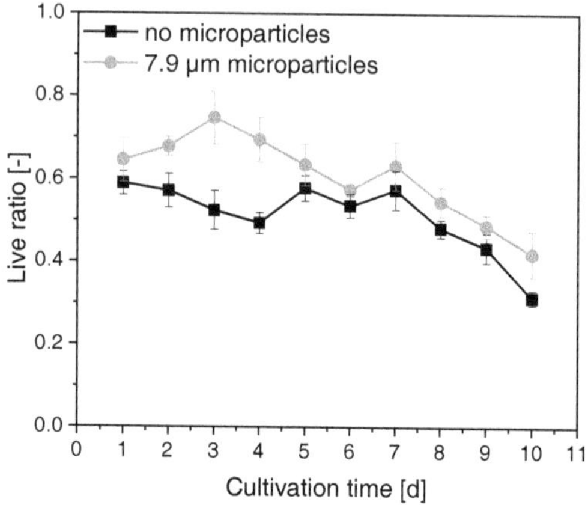

Figure 4.7: Live ratio calculated with equation (12) from fluorescent stained pellet slices from a cultivation without and with 7.9 µm microparticles.

The live ratio of pellets from the cultivation with 7.9 µm microparticles is increased compared to pellets from an unsupplemented cultivation. Especially between day 1 and day 4 of cultivation a significant difference is visible. The live ratio for microparticle treated pellets is increasing until day 3 and is decreasing thereafter, while the live ratio in pellets without microparticle treatment decreases from day 1 to day 4 and then shows a slight increase. During this period *L. aerocolonigenes* is in the exponential growth phase followed by the transition to the stationary phase. The largest difference is observed on day 3. Microparticle treated pellets have a live ratio of approximately 0.75 and untreated pellets of about 0.52 resulting in a difference of 0.23. From day 5 on the course for both approaches is similar with a larger decrease starting after day 7. This decrease is connected to the death phase in which cell lysis leads to a decreased live ratio. After 10 days the live ratio is 0.42 for pellets treated with microparticles and 0.31 for pellets cultivated without microparticles.

The microscopic images of selected stained pellet slices that were used for the calculation of the live ratio are illustrated in **Figure 4.8** in which the same trend can be observed visually. Green areas represent living parts of the pellet, while red areas represent dead parts.

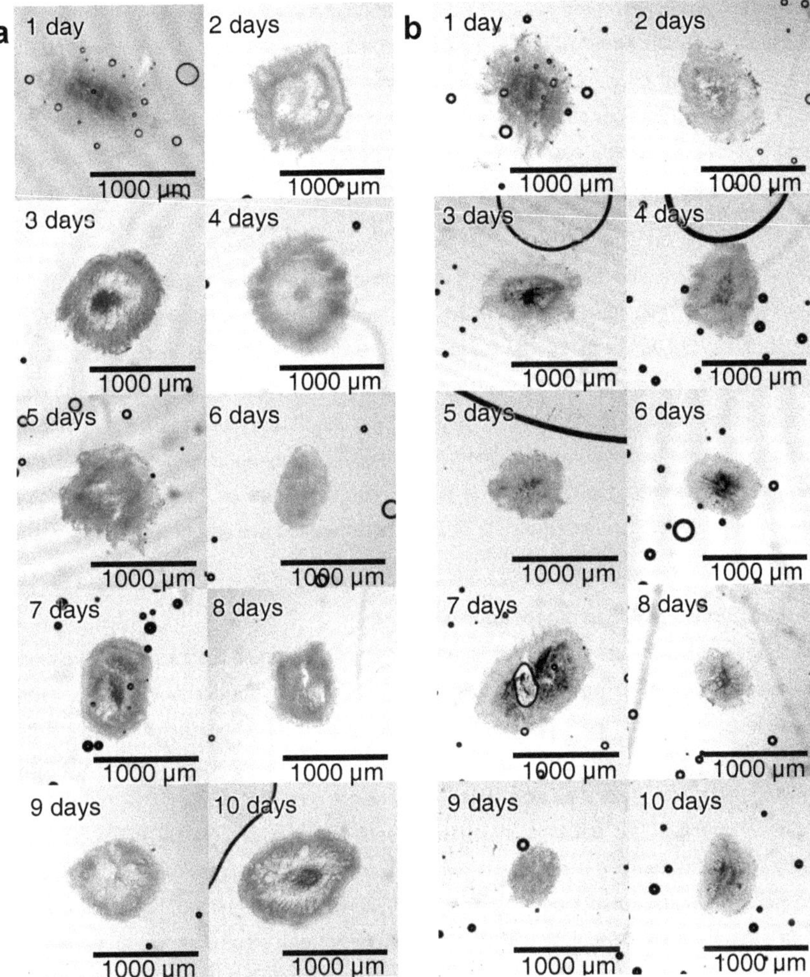

Figure 4.8: Exemplary microscopic images of sliced pellets from (a) an unsupplemented cultivation and (b) a 7.9 µm glass bead supplemented cultivation acquired with a CLSM. The pellets were stained with SYTO9 (green fluorescing) and PI (red fluorescing).

In these images proportionally larger green areas can be seen in the pellets from the microparticle supplemented cultivation. Furthermore, an increasing red area with increasing cultivation time is visible for both approaches.

However, pellet slicing is a time-consuming procedure relying on manual work. The presented results are based on six pellet slices per day and approach. For statistically reliable results, the analysis of larger numbers of pellet slices would be necessary.

Nonetheless, the presented results describe a trend of the effects on viability by microparticle addition. A similar effect was observed by Driouch et al. (2010b), however, not regarding viability but production in slices of pellets or mycelial aggregates. The authors cultivated a GFP producing *A. niger* strain without and with 5 g L^{-1} talc microparticles resulting in pellets and mycelial structures, respectively. While the pellet showed only a thin outer layer of GFP production, the mycelium induced by microparticle addition showed GFP production across the entire mycelium (Driouch et al. 2010b).

Here, the decreased pellet diameter with microparticle addition suggests an improved nutrient and oxygen supply into the pellet, as already described by different authors (Walisko et al. 2012; Driouch et al. 2010a; Gonciarz and Bizukojc 2014). The incorporation effect of microparticles observed in *L. aerocolonigenes* additionally suggests a loosening of the pellet structure by entangled particles in between the single hyphae of a pellet. This might contribute to an improved oxygen and nutrient supply, so that macro-morphological aspects as pellet size are not the only decisive factors. Further studies on the oxygen concentration inside of microparticle treated pellets of *L. aerocolonigenes* are part of a follow-up research project.

4.1.2 Surface modified microparticles

In literature, mostly substances with different chemical properties like talc, aluminum oxide or titanate are applied as microparticles (Antecka et al. 2016a), but due to the inhomogeneous size and shape of these particles the investigation of chemical surface effects of microparticles is impeded. Therefore, different surface modifications of a single type of microparticle can help providing these information. Walisko et al. (2017) already used surface modified talc particles in cultivations of *L. aerocolonigenes*. They observed an increased rebeccamycin concentration when microparticles with a hydrophilic or only slightly hydrophobic surface and a negative zeta-potential were supplemented. However, it was not totally clear if these were the most important properties or if other chemical properties of the surface modification are of intertest for an increased production (Walisko et al. 2017). To investigate the chemical surface effects in more detail, micro glass beads were used for modification in this thesis.

The surface modification of the glass microparticles was done by the iPAT, TU Braunschweig. By binding (3-aminopropyl)triethoxysilane (APTES) to the surface as a linker, the microparticle surface could easily be functionalized with different carboxylic acids. Since the incorporation of microparticles in pellets was mainly observed for the smaller glass microparticles ($x_{50} = 7.9$ µm) applied in this thesis and since this size rather fits most microparticles applied in literature, surface modification was only conducted for this microparticle size. In **Figure 4.9** the cultivation results of a 10 day cultivation with microparticles functionalized with three different carboxylic acids is presented. In this case, the CDW was determined by incineration (compare **chapter 3.4.1**) resulting in values that are not falsified by the microparticles.

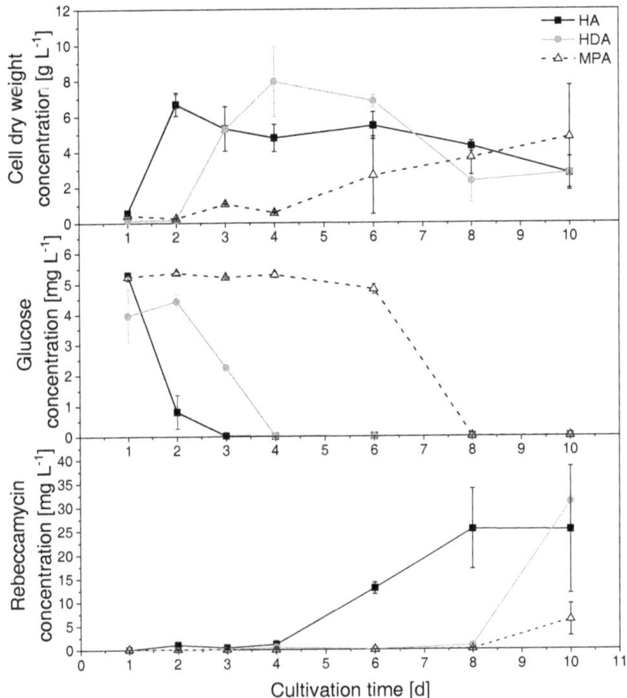

Figure 4.9: Cell dry weight, glucose and rebeccamycin concentrations of a 10 day cultivation with three different surface modified microparticles functionalized with carboxylic acids (HA = hexanoic acid, HDA = hexanedioic acid, MPA = 3-mercaptopropionic acid).

Clear differences in the CDWs of all three approaches can be observed. Addition of microparticles modified with HA show a similar behavior as the addition of unmodified microparticles (compare **Figure 4.5**). The maximal CDW is reached on day 2 followed by a decreasing CDW. Microparticles modified with HDA cause a slight delay in growth. The CDW starts increasing after day 2 and reached the maximum on day 4 of cultivation followed by a decrease. MPA microparticle addition, however, leads to an even further growth delay. Until day 4 only marginal changes in the CDW are observed, a visible increase occurs after 4 days of cultivation. The CDW increases until the end of this cultivation. The glucose concentrations of the three approaches fit with the CDWs. Glucose is fully consumed on day 3 for HA microparticle addition, on day 4 for HDA microparticle addition and on day 8 for MPA microparticle addition. The delayed growth with certain microparticles is also affecting the rebeccamycin concentration. While rebeccamycin production starts after day 4 with HA microparticles, with HDA and MPA microparticles rebeccamycin concentrations increase only after day 8. A slowed down substrate consumption and growth was already observed for the addition of talc particles to a strain of *A. niger*. With microparticle addition fructose and glucose were fully depleted 12 and 18 h later, respectively, compared to an unsupplemented approach. Furthermore, the biomass growth was slowed down and resulted in lower overall concentrations compared to the unsupplemented control (Kowalska et al. 2017). Similar effects are visible in this thesis, with addition of surface modified microparticles to *L. aerocolonigenes*.

Since the only difference between all three approaches is the carboxylic acid bound to the microparticle surface, HDA and MPA seem to be responsible for the growth delay leading to a delayed rebeccamycin production. Except for the carboxyl group needed for the binding to the microparticle surface, HA is only a carbon chain without any additional functional groups. HDA has a second carboxyl group and MPA has a sulfhydryl group (compare **Table 3.2**), making these two substances more reactive. Thus, interactions with the functional groups may cause the differences in glucose, biomass and rebeccamycin concentration. However, the rebeccamycin concentration is usually increasing over a longer time period than two days meaning that a prolonged cultivation would be necessary to compensate for the delayed growth with HDA and MPA microparticles. To further investigate this aspect and the effects of microparticles modified with further carboxylic acids, a broader screening of surface modified microparticles in a prolonged cultivation of 14 days was conducted. Since the morphological changes due to the surface modified microparticles will be considered in detail for this screening, the cell morphology of *L. aerocolonigenes* with addition of HA, HDA and MPA microparticles will not be discussed beforehand. In **Figure 4.10** the CDW and rebeccamycin concentration from a cultivation with microparticles functionalized with six different carboxylic acids (see also **Table 3.2**) and microparticles from different stages in the modification process are presented. The dotted lines are included for a visual separation of control approaches or modification precursors of microparticles and the actual modifications with carboxylic acids.

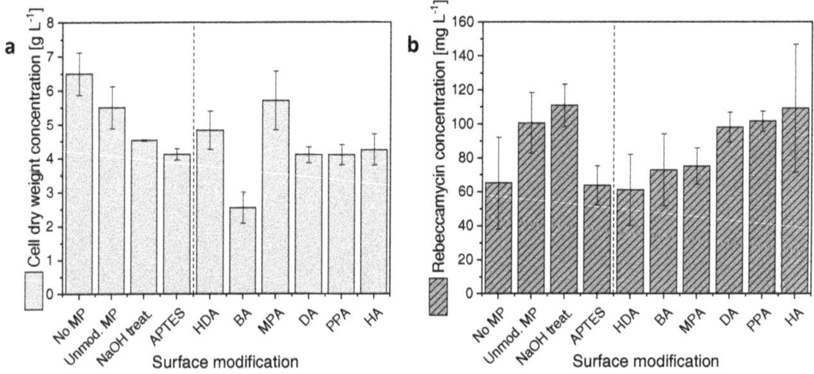

Figure 4.10: (a) Cell dry weight and (b) rebeccamycin concentrations of a 14 day cultivation of *L. aerocolonigenes* with different modified microparticles in a concentration of 10 g L^{-1} (No MP = No microparticles, Unmod. MP = unmodified microparticles, NaOH treat. = microparticles treated with NaOH, APTES = (3-aminopropyl)triethoxysilane, HDA = hexanedioic acid, BA = butanoic acid, MPA = 3-mercaptopropionic acid, DA = decanoic acid, PPA = 3-phenylpropanoic acid, HA = hexanoic acid).

In general, the unsupplemented cultivation (No MP) showed the highest CDW after 14 days of cultivation with around 6.5 g L^{-1} (**Figure 4.10a**). While unmodified microparticles (Unmod. MP) and microparticles modified with HDA and MPA led to only slightly lower CDWs, the addition of microparticles modified with BA clearly resulted in the lowest CDW with around 2.5 g L^{-1}. Considering the rebeccamycin concentration (**Figure 4.10b**), the addition of unmodified microparticles or microparticles treated with NaOH led to an increased titer with about 101 or 111 mg L^{-1}, respectively, compared to the unsupplemented control with 65 mg L^{-1} rebeccamycin. The addition of microparticles functionalized with only the linker molecule APTES resulted in a similar rebeccamycin concentration as the unsupplemented control. Microparticles functionalized with different carboxylic acids had different effects on the rebeccamycin concentration. Some modified microparticles (HDA, BA, MPA) led to similar rebeccamycin concentrations as in the unsupplemented control, but some also increased the rebeccamycin titer (DA, PPA, HA). The largest rebeccamycin concentrations were achieved with DA (98 mg L^{-1}), PPA (102 mg L^{-1}) and HA (109 mg L^{-1}). Although these values are similar to the rebeccamycin titers of unmodified and NaOH treated microparticle addition, none of the applied surface modified microparticles led to rebeccamycin concentrations higher than this. These results imply that the microparticle modifications applied are either disadvantageous for the microorganisms or are not influencing the cultivation process behaving similar to unmodified microparticles.

Since microparticles functionalized with different carboxylic acid had different effects on the CDW and the rebeccamycin concentration, the effects on the pellet morphology of *L. aerocolonigenes* was also investigated (**Figure 4.11**).

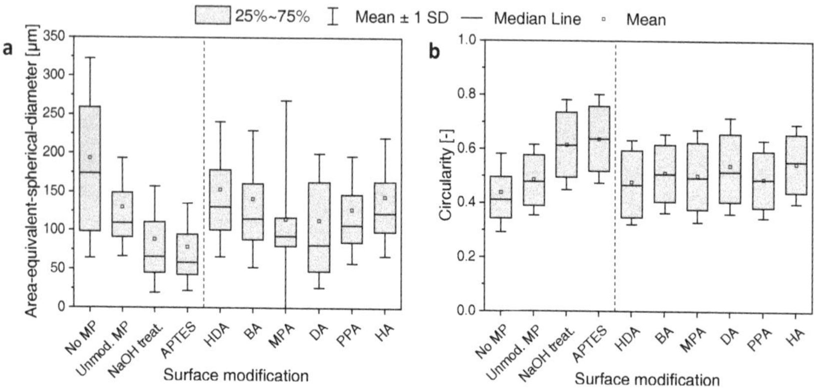

Figure 4.11: (a) Area-equivalent-spherical diameter and (b) circularity of pellets grown with different surface modified microparticles in a concentration of 10 g L^{-1} (No MP = No microparticles, Unmod. MP = unmodified microparticles, NaOH treat. = microparticles treated with NaOH, APTES = (3-aminopropyl)triethoxysilane, HDA = hexanedioic acid, BA = butanoic acid, MPA = 3-mercaptopropionic acid, DA = decanoic acid, PPA = 3-phenylpropanoic acid, HA = hexanoic acid).

The AESD (**Figure 4.11a**) of the approach without microparticles has the widest distribution with the highest mean and median value. Hence, all microparticles added to *L. aerocolonigenes* led to smaller pellets. The smallest pellets were achieved with addition of NaOH treated and APTES microparticles. All others are located in between. For the addition of MPA microparticles a rather large standard deviation despite a narrow distribution can be observed. In the microscopic images many small filaments with only a few very large pellets were visible. Representative microscopic images of the cultures are presented in **Figure 4.12**. In case of the circularity (**Figure 4.11b**) the unsupplemented approach led to the smallest mean values compared to all microparticle supplementations. The highest circularities were achieved with the addition of NaOH treated microparticles or microparticles modified with APTES. For microparticles functionalized with different carboxylic acids no great differences were visible. Although general differences in the AESD and the circularity of *L. aerocolonigenes* with addition of different microparticles were observed, no clear correlation to the resulting rebeccamycin concentrations can be made. A general decrease in pellet size and increase in circularity with microparticle addition might play a role, but the morphology does not seem to be the main driving force for the differences in the rebeccamycin titers.

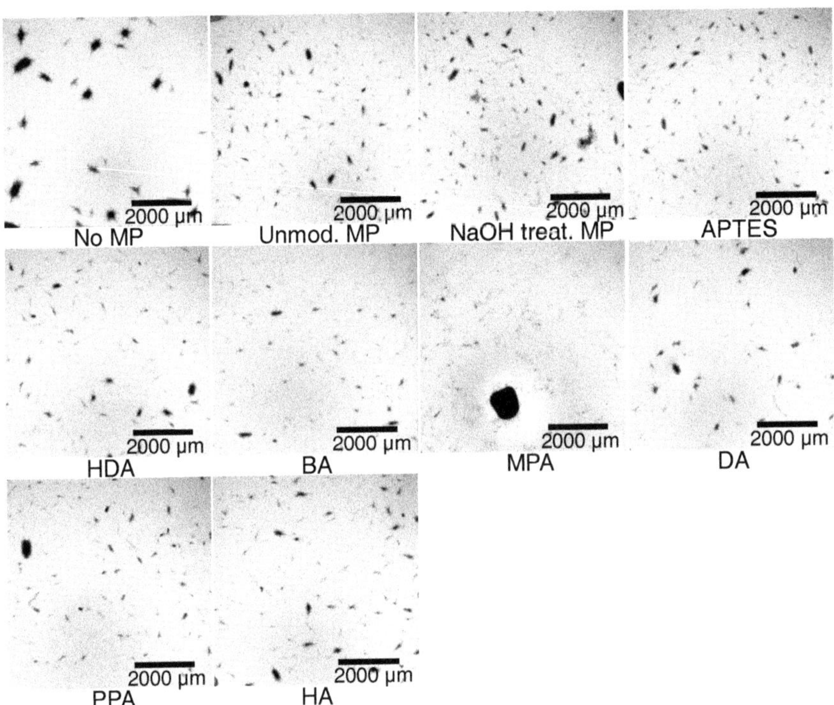

Figure 4.12: Microscopic images of *L. aerocolonigenes* under addition of different surface modified microparticles in a concentration of 10 g L^{-1} (No MP = No microparticles, Unmod. MP = unmodified microparticles, NaOH treat. = microparticles treated with NaOH, APTES = (3-aminopropyl)triethoxysilane, HDA = hexanedioic acid, BA = butanoic acid, MPA = 3-mercaptopropionic acid, DA = decanoic acid, PPA = 3-phenylpropanoic acid, HA = hexanoic acid).

As the cell macro-morphology alone does not give an explanation for the different effects of different surface modified microparticles, the inner pellet structure should be examined in more detail. The incorporation of microparticles into pellets of *L. aerocolonigenes* has already been observed before (compare **Figure 4.4**), therefore this aspect was also considered for the addition of surface modified microparticles. In **Figure 4.13** pellets slices of all presented approaches are displayed.

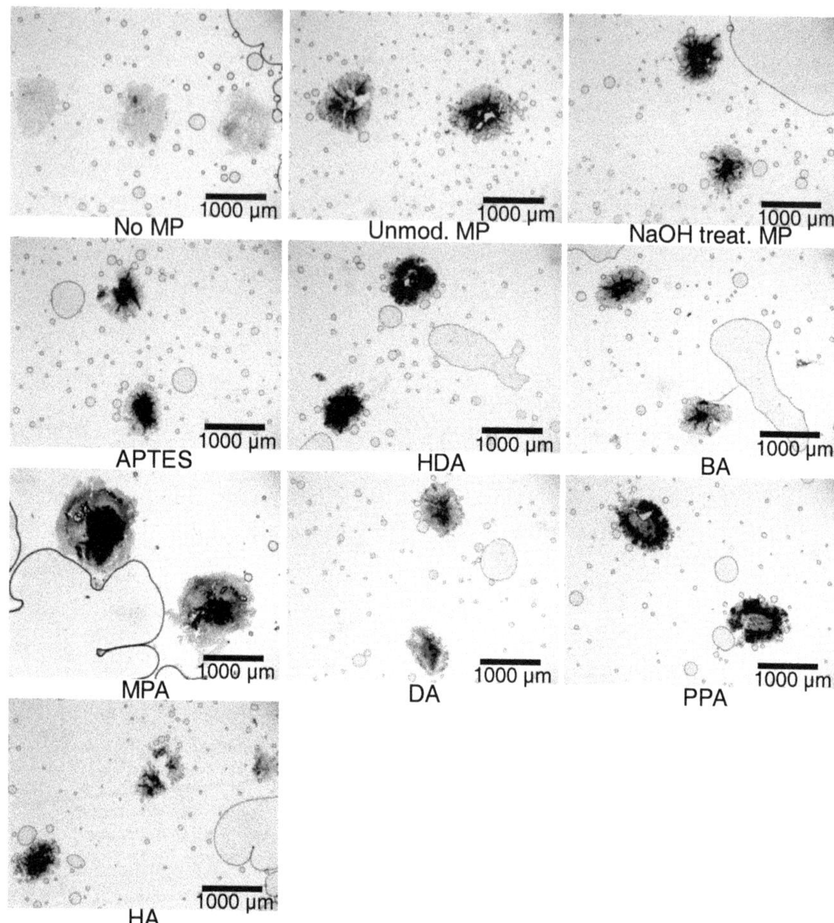

Figure 4.13: Microscopic images of pellet slices from cultivations with different surface modified microparticles in a concentration of 10 g L^{-1}. Black lines or circles are air bubbles occasionally enclosed in the sectioning medium (No MP = No microparticles, Unmod. MP = unmodified microparticles, NaOH treat. = microparticles treated with NaOH, APTES = (3-aminopropyl)triethoxysilane, HDA = hexanedioic acid, BA = butanoic acid, MPA = 3-mercaptopropionic acid, DA = decanoic acid, PPA = 3-phenylpropanoic acid, HA = hexanoic acid).

In all microparticle supplemented approaches the particles were incorporated into pellets. Some pellet slices contain fractures or holes that may be caused by the slicing process of the fragile pellets. The microscopic image from the unsupplemented approach shows pellet slices of an untreated pellet for comparison. For the different surface modified microparticles different patterns for the incorporation can be observed. Unmodified, NaOH treated, APTES, BA, DA and HA microparticles lead to a similar incorporation pattern with a star shaped accumulation of microparticles, in some cases with a free core area. However, HDA microparticles seem to be homogenously distributed throughout the pellet. MPA microparticles form a dense particle core within the pellets and the outer pellet area is free of particles. PPA particles led to an opposite pattern. No microparticles are present in the pellet core, but they accumulate in a circular pattern around the core. The special incorporation patterns occur with supplementation of microparticles modified with the carboxylic acids with an additional functional group or an aromatic structure. This implies that these functional group could play a major role regarding the type of incorporation of microparticles.

Microparticle incorporation or especially incorporation patterns developing in combination with microparticle-enhanced cultivation are only rarely described in literature. Walisko et al. (2017) noticed particle entanglement in pellets of *L. aerocolonigenes* when talc microparticles were added. Similar observation were made by Kuhl et al. (2020) for the addition of talc particles to *S. albus*. Driouch et al. (2012) even described core-shell pellets of *A. niger* with titanate addition in which the microparticles formed a dense core. A similar pattern is visible for the addition of MPA microparticles to *L. aerocolonigenes*. In some cases microparticle addition led to freely dispersed mycelium, where these types of incorporation patterns seem to be unlikely (e.g. Kaup et al. (2008) and Driouch et al. (2010a)). However, none of the other studies cited in this thesis described microparticle incorporation.

Although no clear correlation between particle properties, incorporation pattern and rebeccamycin titer is found, the approach of surface modified microparticles seems purposeful. As Etschmann et al. (2015) already proposed, surface modified microparticles are similar in size and shape and only vary in their surface properties allowing a better comparison of physicochemical interactions of microparticles and hyphae. The many recently published studies on microparticle-enhanced cultivation (compare **Table 2.2**) indicate that this topic is highly researched. Nevertheless, a clear mechanism has not yet been identified. The here applied surface modified microparticles led to a similar or increased rebeccamycin titer as an unsupplemented approach. However, the rebeccamycin titer with unmodified microparticles could not be exceeded. Incorporation of microparticles could play a major role in the productivity increase, as the hyphal growth is changed and might lead to a looser pellet structure resulting in an increased nutrient and oxygen supply. Further surface modifications need to be investigated to shed light on this topic.

4.2 Macroparticle enhanced cultivation

A not yet fully investigated approach for particle enhanced cultivations of filamentous microorganisms is the addition of larger macroparticles (Ø > 200 µm). Often chemically inert glass particles are used for this purpose (Dobson et al. 2008; Holtmann et al. 2017). This method already proved successful for *L. aerocolonigenes* (Walisko et al. 2017; Schrader et al. 2019) and further investigations regarding this topic are described and discussed in this chapter.

4.2.1 Screening of particle diameters

Walisko et al. (2017) investigated the effect of different glass bead diameters (0.5, 1 and 2 mm) on cultivations of *L. aerocolonigenes* at a fixed concentration of 80 g L^{-1}. Highest rebeccamycin concentrations were achieved with 0.5 mm glass beads using another shake flask geometry than applied in this thesis. In Walisko et al. (2017) the baffles were located at the bottom of the shake flask while the baffles were located on the shake flask sides in this thesis (see Schrader et al. (2019) for details). Since particle size seems to be an important factor, eight diameters between 250 and 2000 µm were chosen for further investigations. The rebeccamycin concentrations resulting from a cultivation with these glass bead diameters from two different biological approaches are illustrated in **Figure 4.14a** and **Figure 4.14b**.

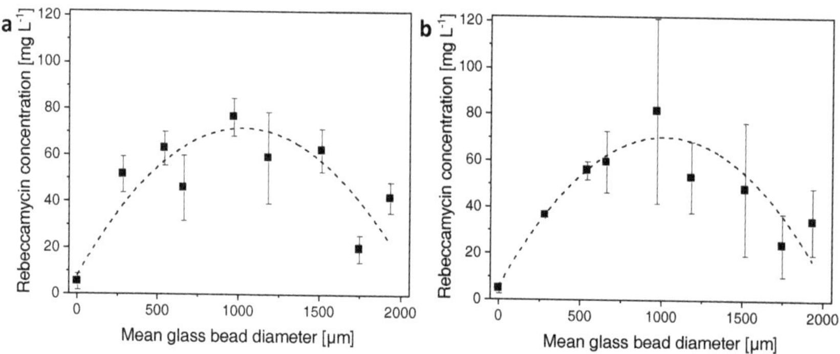

Figure 4.14: Rebeccamycin concentrations of a 10 day cultivation with different glass bead diameters at a concentration of 100 g L^{-1}. The same cultivation was conducted with different biological approaches in (a) and (b).

Although not totally clear, a slight trend of an increasing rebeccamycin concentration with increasing mean glass bead diameter up to a certain diameter (969 µm) and a decrease with further increasing diameters is observed. As already indicated above, the rather large standard deviations and fluctuations of the data points lead to a lack of clarity in these results. This is the reason why the results of two different biological approaches of the same cultivation are presented. Despite these disadvantages the course of the rebeccamycin concentration depending on the mean glass bead diameter is qualitatively similar.

Compared to a control without glass beads (depicted at 0 µm) with a rebeccamycin concentration of around 5 mg L^{-1} in both approaches the addition of 100 g L^{-1} of 969 µm glass beads lead to 77 and 82 mg L^{-1} rebeccamycin, respectively. Since the glass beads used for supplementation are biologically and chemically inert, the interaction between *L. aerocolonigenes* and the glass beads is purely mechanical. The stress energy describes the maximal amount of energy that can be transferred from a glass bead to a pellet in a single stress event (Kwade 2003). How much stress energy is induced from the glass beads onto the microorganism depends, amongst others, on the particle size (equation (1)). With increasing glass bead diameter, the induced stress energy increases (if they meet with the same relative speed). Combined with the trend observed in **Figure 4.14**, this suggests that a certain amount of mechanical stress is necessary to achieve optimal rebeccamycin concentrations. Larger mechanical stress (i.e. induced by larger particles) is unfavorable and at some point, leads to the destruction of *L. aerocolonigenes* pellets while lower mechanical stress is not sufficient. To further understand the effects of glass bead addition in cultivations of *L. aerocolonigenes* more detailed investigations were conducted.

4.2.2 Kinetic and physiological effects of glass bead addition

In the cultivation with different glass bead diameters described above only final rebeccamycin concentrations were considered. To find out about the influence of glass beads during growth or product formation, the time course of the cultivation needs to be investigated in detail. Thus, a cultivation approach without and with 100 g L^{-1} of 969 µm glass beads were compared over cultivation time with daily sampling (**Figure 4.15**). All results related to this cultivation were part of a cooperation with the iPAT (TU Braunschweig), the TU Wien (flow cytometry) and the TU München (morphology) (Schrinner et al. 2020).

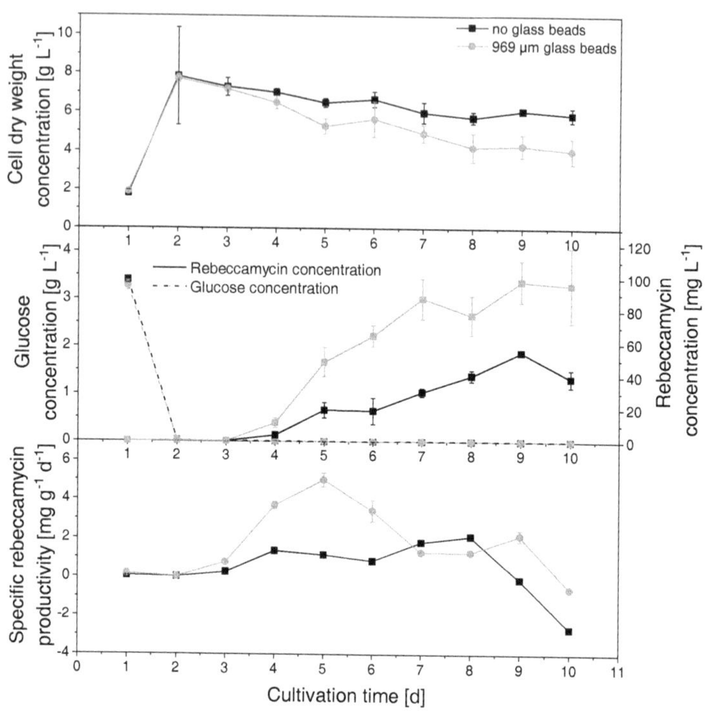

Figure 4.15: Cell dry weight, glucose and rebeccamycin concentrations as well as the specific rebeccamycin productivity of a 10 day cultivation without and with 100 g L^{-1} of 969 µm glass beads.

The CDWs of both approaches are increasing until day 2 of cultivation to a similar value of about 8 g L^{-1} and decrease thereinafter. The CDW decrease is slightly faster for the glass bead supplemented cultivation than in the unsupplemented approach. The glucose consumption is also similar for both approaches, being nearly fully depleted after 2 days of cultivation. The similar course of glucose and CDW concentrations indicate that glass bead addition does not directly affect the growth kinetics of *L. aerocolonigenes*. Growth is neither accelerated nor slowed down. Furthermore, the similar CDW concentration especially in the beginning of the cultivation shows that no biomass dependent increase in rebeccamycin occurred. Rebeccamycin formation starts between day 3 and 4 of cultivation for both approaches and is increasing much faster for the glass bead supplemented cultivation. After 10 days without particles almost 39 mg L^{-1} and with particle supplementation around 95 mg L^{-1} of rebeccamycin were produced. The biomass specific rebeccamycin productivity q_P is increased for the glass bead supplemented cultivation compared to the unsupplemented cultivation, especially in the exponential production phase between 3 to 6 days. The maximal specific rebeccamycin productivity with glass beads was 5.01 mg g^{-1} d^{-1} at day 5 of cultivation while at the same time the specific productivity without glass beads was only 1.14 mg g^{-1} d^{-1}. The maximal specific productivity in the unsupplemented appeared on day 8 of cultivation with 2.10 mg g^{-1} d^{-1}.

To further investigate the impact of glass bead addition to *L. aerocolonigenes* the viability of the pellets was investigated. Viability investigations based on pellet slices and fluorescent staining was already used for the addition of microparticles to *L. aerocolonigenes* (see **chapter 4.1.1**). However, the low number of pellets that can be examined with this method results in a certain lack of reliability of the data. Therefore, flow cytometry with additional fluorescent staining was applied for the investigation of glass macroparticle effects, allowing larger numbers of at least 50 pellets per sample to be examined. The results from flow cytometry measurements are presented in **Figure 4.16**.

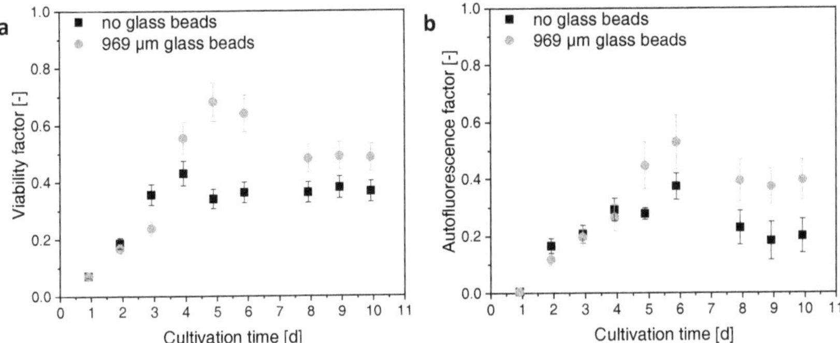

Figure 4.16: (a) Viability factor (calculated with equation (16)) and (b) autofluorescence factor (calculated with equation (17)) of *L. aerocolonigenes* pellets treated without and with 100 g L^{-1} of 969 µm glass beads during a 10 day cultivation originating from flow cytometry measurements.

The viability factor (equation (16)) shown in **Figure 4.16a** gives information about the pellet viability over time. In the beginning the viability factor increases similarly for the cultivations without and with glass beads. From day 4 to 6 a significant difference between both approaches is visible. The viability factor of pellets from a glass bead supplemented cultivation are increased compared to untreated pellets. In the same time period, the exponential rebeccamycin production occurs (see **Figure 4.15**) with a significantly increased specific productivity of the glass bead supplemented approach. On day 5 the viability factor for the glass bead treated pellets was around 0.68 while untreated pellets only achieved a value of 0.34. Since FDA was used for fluorescent staining, conclusions on the metabolic activity of the investigated pellets are possible. The presented results suggest an increased metabolic activity of glass bead treated pellets in this period leading to an increased production (Leštan et al. 1996; Li et al. 2011; Saruyama et al. 2013). A similar beneficial effect of larger active, growing zones was also observed during the production of the secondary metabolite lovastatin in *A. terreus* (Bizukojc and Ledakowicz 2010). Furthermore, the autofluorescence factor (**Figure 4.16b,** equation (17)) was determined for which the measurements were conducted without fluorescent staining. The course over time is similar to the viability factor but with lower total values. Here again, an increased autofluorescence factor of pellets from a glass bead supplemented cultivation especially in the period of exponential production was observed. Hence, the connection to an increased productivity is proposed, because rebeccamycin might act as a fluorescent substance itself (Moreau et al. 1999).

4.2.3 Morphological pellet characterization with glass bead addition

Effects of glass bead addition on the morphology of filamentous *L. aerocolonigenes* were not yet considered. For many filamentous microorganisms changes in cell morphology are linked to an increased production and should therefore be considered (Walisko et al. 2015).

The pellet macro-morphology in the cultivation described in the previous chapter was investigated via microscopy and flow cytometry to find out about the effects of glass bead addition on morphology. In **Figure 4.17** the pellet size results of microscopy followed by image analysis are displayed. Microscopic images were acquired every day during the 10 day cultivation resulting in the analysis of approximately 158,600 pellets, with 54,800 pellets from the cultivation without glass beads and 103,800 pellets from the glass bead supplemented cultivation.

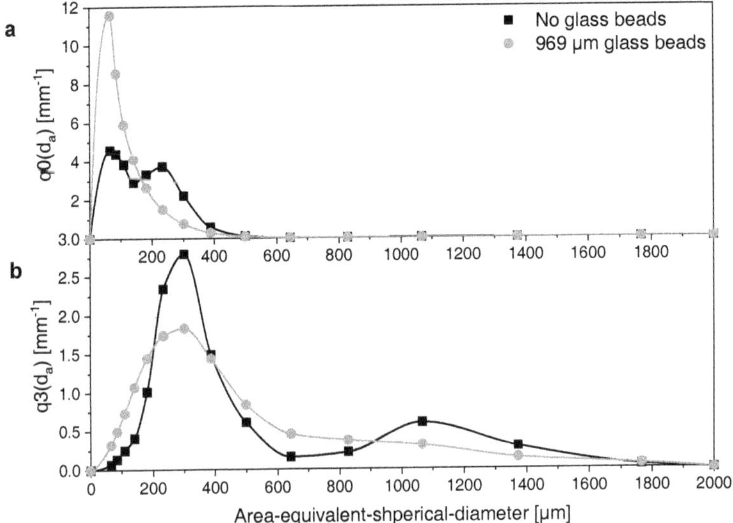

Figure 4.17: Normalized (a) number-density-distribution $q_0(d_a)$ and (b) volume-density-distribution $q_3(d_a)$ of the area-equivalent-spherical-diameter obtained by microscopic image analysis of pellets from a ten day cultivation without or with glass beads.

The number-density-distributions of the AESD (**Figure 4.17a**) of all pellets analyzed during the 10 day cultivation indicate differences in pellet size with and without glass bead treatment. Pellets from the glass bead supplemented cultivation are smaller; the mean AESD of these pellets is 106 µm and for pellets from the unsupplemented cultivation 172 µm. Most pellets are in the size range between 0 and 500 µm. Looking at the volume-density-distributions (**Figure 4.17b**) few larger pellets become apparent, especially between 1000 µm and 1200 µm for the cultivation without glass beads. Further morphological parameters, the Feret diameter and the circularity, were investigated and are presented in **Figure 4.18**.

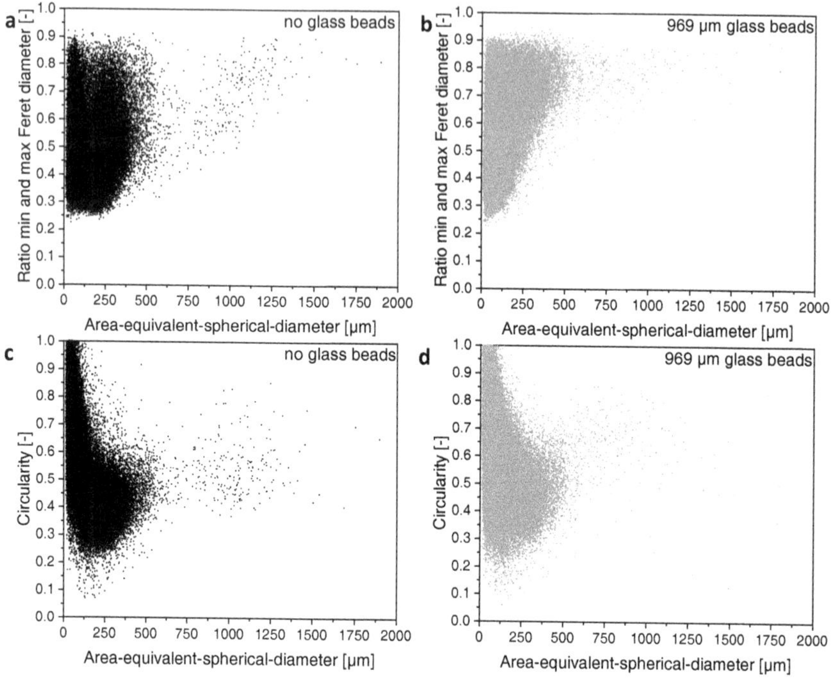

Figure 4.18: Morphological parameters of pellets from a cultivation without and with 100 g L^{-1} of 969 µm glass beads. Ratio of minimum and maximum Feret diameter and area-equivalent-spherical-diameter for pellets (a) without and (b) with glass beads, circularity and area-equivalent-spherical-diameter (c) without and (d) with glass beads are displayed.

For a deeper insight into the cell morphological effects of glass bead addition, the ratio of minimum and maximum Feret diameter and the circularity were calculated. Both parameters are measures for the roundness of pellets. The ratio of minimum to maximum Feret diameter of the pellets has a mean value of 0.50 ± 0.07 in the unsupplemented cultivation (**Figure 4.18a**) and 0.58 ± 0.07 in the cultivation with glass beads (**Figure 4.18b**). The mean of the circularity of pellets from the cultivation without glass beads is 0.51 ± 0.09 (**Figure 4.18c**) and 0.64 ± 0.09 for glass bead treated pellets (**Figure 4.18d**). The results indicate that the pellets from the glass bead supplemented cultivation are more circular than pellets from the unsupplemented cultivation. This can also be observed by the distribution of the single data points in the scatter plot (**Figure 4.18**). In general, the morphological analysis of microscopic images showed smaller and more circular pellets when glass beads were added. These are similar to the results of sample length and compactness from macro-morphological analysis via flow cytometry (**Figure 4.19**).

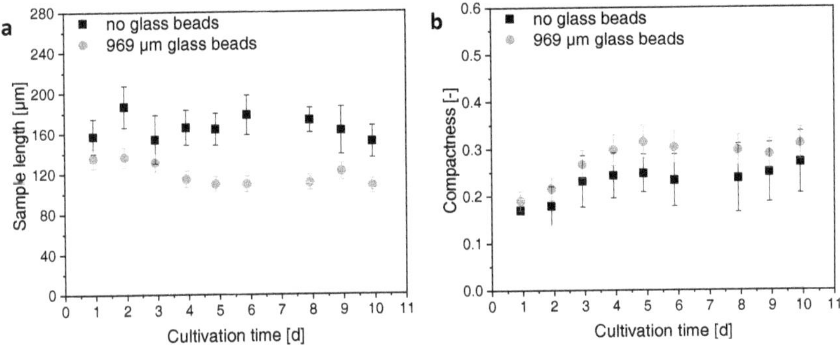

Figure 4.19: (a) Sample length and (b) compactness (calculated with equation (13)) of pellets from a 10 day cultivation of *L. aerocolonigenes* without and with 100 g L^{-1} of 969 µm glass beads obtained by flow cytometry.

The sample length, describing the pellet diameter, is generally lower for the glass bead supplemented cultivation (**Figure 4.19a**). The overall mean sample length is 100 µm for glass bead treated pellets and 170 µm for pellets from an unsupplemented cultivation. A further morphological parameter is the compactness (equation (13)) which is slightly increased for glass bead treated pellets (**Figure 4.19b**).

The results from both microscopic and flow cytometric analysis indicate that glass bead addition leads to smaller, more uniform and rounder pellets. For the addition of glass beads to *L. aerocolonigenes* smaller pellets were also observed in a previous study (Walisko et al. 2017). An increased productivity in smaller pellets of filamentous microorganisms is often linked to an improved substrate supply, especially oxygen, since large pellets often exhibit substrate limitations in the pellet core (Wucherpfennig et al. 2010). The penetration depth of oxygen in pellets was investigated via microelectrodes for several filamentous fungi (Wittler et al. 1986; Cronenberg et al. 1994; Hille et al. 2005, 2009; Bizukojc and Gonciarz 2015). Hille et al. (2005) investigated the oxygen transfer in pellets of *A. niger*. The measured oxygen concentration profiles inside small pellets with a diameter around 400 µm showed a sufficient oxygen supply whereas larger pellet diameters of 800 µm and higher resulted in oxygen limitations in the pellet core (Hille et al. 2005). Similar results were obtained for microelectrode measurements in *A. terreus* (Bizukojc and Gonciarz 2015) and *P. chrysogenum* (Cronenberg et al. 1994).

However, to get a full insight into the morphological changes by glass bead addition, the analysis of a single cultivation approach is not sufficient. In Schrader et al. (2019) an increased pellet diameter of *L. aerocolonigenes* with glass bead addition compared to an unsupplemented cultivation was observed under the same conditions. Walisko (2017) observed smaller pellets, but a decreased circularity. At first sight these results seem to be contradictory, therefore, the pellet macro-morphology was investigated in several further cultivations with glass bead addition (**Figure 4.20**). The macro-morphology was analyzed only on day 10 of the cultivation; in each case an unsupplemented approach and an approach with 100 g L^{-1} of 969 µm glass beads were cultivated together.

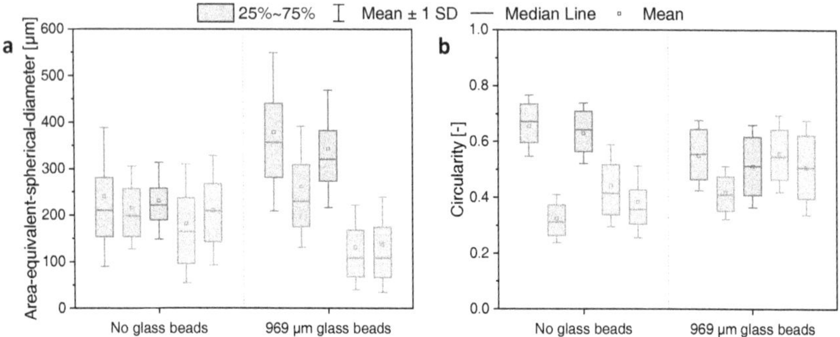

Figure 4.20: (a) Area-equivalent-spherical-diameter and (b) circularity of pellets from cultivations without or with 100 g L^{-1} of 969 µm glass beads. Each color represents a different biological approach.

The AESD of pellets cultivated without glass beads are in a similar area for different cultivation approaches whereas for pellets cultivated with glass beads the AESD is fluctuating between different cultivations (**Figure 4.20a**). However, an opposite effect is observed for circularity (**Figure 4.20b**). Without glass bead treatment the circularity of pellets is varying considerably in different approaches while the circularity seems more regular with glass bead addition. Furthermore, no clear trend of increased or decreased diameter or circularity with glass bead addition is observed.

This indicates, that the effect of glass beads on the pellet size is rather indistinct and most likely not the reason for an increased rebeccamycin productivity. A stable circularity caused by glass bead addition might be relevant for an increased productivity, but it is probably not the only reason. Tesche et al. (2019) observed an increased concentration of labyrinthopeptin A1 in a salt-enhanced cultivation of *Actinomadura namibiensis* leading to an increased circularity of pellets. Despite the connection of morphology and productivity described for many filamentous microorganisms (Dobson et al. 2008; Tamura et al. 1997; Vecht-Lifshitz et al. 1992; Yin et al. 2008), there are also examples in literature stating otherwise. For the production of clavulanic acid in *S. clavuligerus* and virginiamycin in *S. virginiae* similar productivities were observed in different morphological forms (Belmar-Beiny and Thomas 1991; Yang et al. 1996). This might also apply for *L. aerocolonigenes*. Another conceivable reason for a productivity increase induced by glass bead addition are changes in the micro-morphology, that were not yet characterized. Wardell et al. (2002)

described an influence of the micro-morphological branching rate in *Saccharopolyspora erythraea*. Mutants with a reduced branching rate showed an enhanced hyphal strength. These mutants produced larger amounts of erythromycin than the original strain. The biomass growth was similar for all approaches (Wardell et al. 2002). Moreover, other morphological effects regarding the inner pellet structure induced by glass bead addition, such as an increased pellet porosity, are possible. In this way the oxygen transport into the pellet might be facilitated leading to a better oxygen availability and therefore a better production performance. Hille et al. (2005) also considered the hyphal fraction in radial coordinate of the pellet in their study on oxygen transfer in pellets of *A. niger*. A sufficient oxygen supply was not only dependent on pellet size but also on the radial biomass density inside a pellet (Hille et al. 2005). Of course, there is still the possibility of a morphology unrelated reason for the productivity increase by glass beads, for example a stimulation of the metabolic activity by the right amount of mechanical stress.

4.2.4 Variation of stress energy and stress frequency

As already described above, the stress energy input can be varied e.g., by different particle diameters. However, not only the stress energy but also the stress frequency is important for the overall induced mechanical stress by glass beads. The stress frequency describes the number of stress events during a certain time period (Kwade 2003). An approximation of these two parameters is described in equations (1) - (3), in which the specific energy input is proportional to the product of stress energy and stress frequency. However, this model is only an approximation for the here described approach since it was developed for the characterization of disintegration effects in stirred media mills (Kwade 2003). For an appropriate estimation of the mechanical stress induced by glass beads in shaking flasks, CFD-DEM-simulations are necessary to depict this special process. An insight into these simulations is given in Schrader et al. (2019). Nevertheless, the given equations are a good indicator of parameters that can influence the induced mechanical stress.

The stress energy can be varied by different particle diameters, different particle densities or a variation of the shaking speed while the stress frequency can be varied by a different particle numbers in the shaking flask (see equations (1) - (3)). In a first approach the stress frequency is varied. Therefore, a single particle size was chosen and different particle concentrations (or particle numbers) of this particle diameter were added to a cultivation of *L. aerocolonigenes* (**Figure 4.21**). A mean glass bead diameter of 969 µm was chosen for this investigation as it generally showed to be beneficial in terms of production.

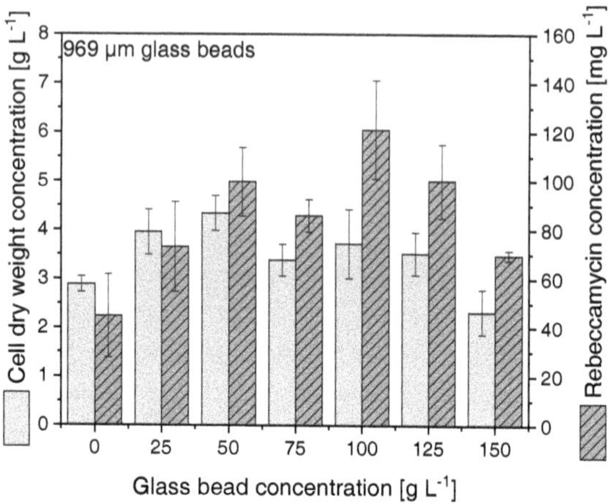

Figure 4.21: Cell dry weight and rebeccamycin concentrations of a 10 day cultivation with different concentrations (or numbers) of 969 µm glass beads.

At first the CDW is increasing with increasing particle concentration with maximal 4.3 g L^{-1} at 50 g L^{-1} glass beads. Hereinafter the CDW is rather constant and decreasing at high concentrations. The rebeccamycin concentration is increasing until a glass bead concentration of 100 g L^{-1} with around 121 mg L^{-1} and is decreasing for 125 and 150 g L^{-1} glass beads. Without particle addition a rebeccamycin titer of only 45 mg L^{-1} was achieved.

In this case, the stress frequency is increased with increasing glass bead concentration. At first the rebeccamycin concentration is increasing correspondingly, however, after a certain point the stress frequency is probably too high leading to a large specific energy being induced. After this point the rebeccamycin concentration decreases (125 g L^{-1} glass beads). The clearly lower CDW at a concentration of 150 g L^{-1} glass beads, at which the rebeccamycin concentration is already decreased, suggests the occurrence of cell destruction.

The variation of the specific energy input by adding glass particles of different diameters was already discussed in chapter **4.2.1**. However, in this case not only the diameter was varied but also the particle number since the same weight concentration was used for each size. In the following approach the effect of the same number of particles for different particle sizes was investigated (**Figure 4.22**). 4200 particles per shaking flask were used for each particle diameter. The weight of this particle number was calculated from particle density and size for each particle diameter and added to the shaking flask.

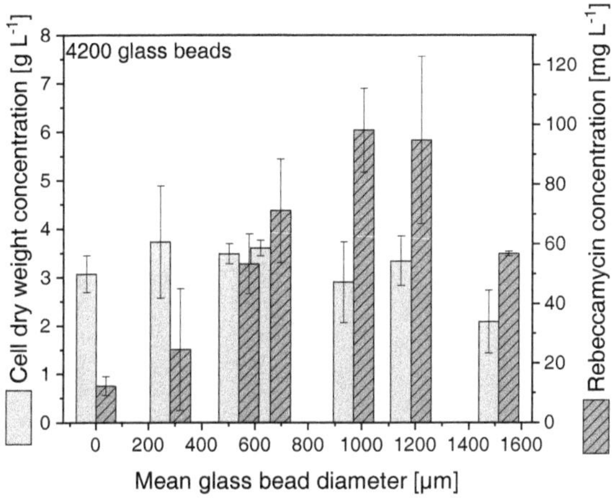

Figure 4.22: Cell dry weight and rebeccamycin concentrations of a 10 day cultivation of different glass bead sizes with the same particle number of 4200.

Here, only particles up to a mean diameter of 1513 µm were used since the added particles already took up a very large volume of the shaking flasks at this point. The CDW does not show a clear trend and mostly fluctuates around 3.5 g L^{-1}. The rebeccamycin concentration increases until a mean glass bead diameter of 969 µm and decreases afterwards, a similar course as observed in **Figure 4.14**. However, in this case clearer differences between the rebeccamycin concentrations of approaches with different glass bead diameters are visible.

The stress energy is increased with increasing glass bead diameter. Since the glass bead number is the same in all approaches, the differences are mainly caused by changes in the stress energy. In equation (1) the particle diameter influences the stress energy to the power of three, meaning that the glass bead diameter is a large influencing factor. The addition of glass beads with a mean diameter of 282 µm resulted in about 24 mg L^{-1} rebeccamycin while the same number of 969 µm glass beads led to around 98 mg L^{-1}. Hence, the alteration of the induced specific energy by different diameters only, greatly affects rebeccamycin titers.

In a further approach the particle density was varied by adding ceramic beads instead of glass beads. The glass beads had a density of 2500 kg m^{-3} while the ceramic beads had a density of 3800 kg m^{-3}. A mean ceramic bead diameter of 918 µm was used in different concentrations between 0 and 100 g L^{-1} (**Figure 4.23**). This particle size was chosen due to the similar size of the most applied glass beads with a mean diameter of 969 µm.

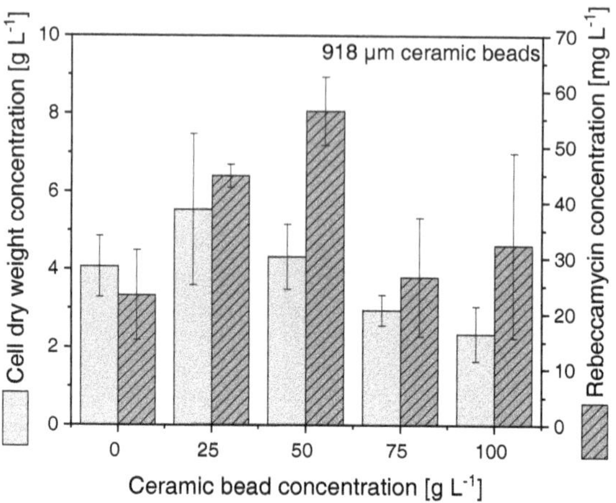

Figure 4.23: Cell dry weight and rebeccamycin concentrations of a 10 day cultivation with addition of different concentrations of 918 µm ceramic beads.

The CDW is decreasing with increasing ceramic bead concentration. The rebeccamycin concentration is increasing until a ceramic bead concentration of 50 g L^{-1} with 56 mg L^{-1} rebeccamycin and is decreasing hereinafter. For the glass bead addition, a concentration of 100 g L^{-1} of 969 µm beads indicated to be most beneficial in regard to product titers. For ceramic beads the optimal concentration of similarly sized beads is only half as large. However, when these two conditions are compared not only the particle density but also the particle number is different. In this case both, stress energy and stress frequency are influenced. For the same particle number of ceramic beads as are present in 100 g L^{-1} glass beads, a concentration of approximately 130 g L^{-1} would be necessary. However, since the rebeccamycin concentration is already decreasing from 75 g L^{-1} on, even lower rebeccamycin concentrations would be expected for 130 g L^{-1} due to an increased specific energy input and thus increased cell destruction. Hence, increased stress energy by an increased particle density needs to be compensated by the reduction of another parameter, for example the particle number, for optimal rebeccamycin titers.

A further possibility for the stress energy variation is the application of different shaker frequencies. Usually, cultivations of *L. aerocolonigenes* in shake flasks were conducted at a shaking frequency of 120 min^{-1}. To investigate the effect of the shaking frequency with glass bead addition cultivations with the different shaking frequencies of 100 and 160 min^{-1} were performed. Since the variation of the shaking frequency also influences the oxygen input into the cultivation broth, the dissolved oxygen (DO) in the broth was monitored to watch out for possible oxygen limitations. The DO in cultivations without glass beads and with shaking frequencies of 100, 120 and 160 min^{-1} is presented in **Figure 4.24**.

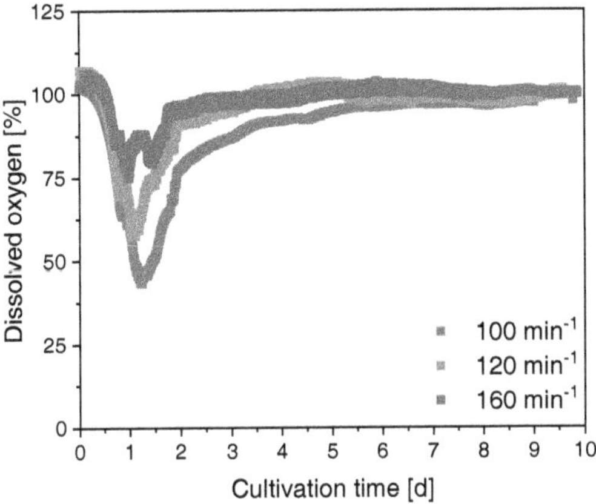

Figure 4.24: Dissolved oxygen in cultivations of *L. aerocolonigenes* without glass beads at different shaking frequencies of 100, 120 and 160 min^{-1}.

The minimal DO values are 75 % for 160 min^{-1}, 55 % for 120 min^{-1} and 43 % for 100 min^{-1} during the growth of *L. aerocolonigenes*. Hence, no oxygen limitation can be observed for any of the applied shaking frequencies. This means that the changes at different shaking frequencies are only caused by a variation of the induced mechanical stress.

The CDWs and rebeccamycin concentrations of cultivations with 100 and 160 min^{-1} are displayed in **Figure 4.25**. With each frequency different glass bead diameters at a concentration of 100 g L^{-1} were used.

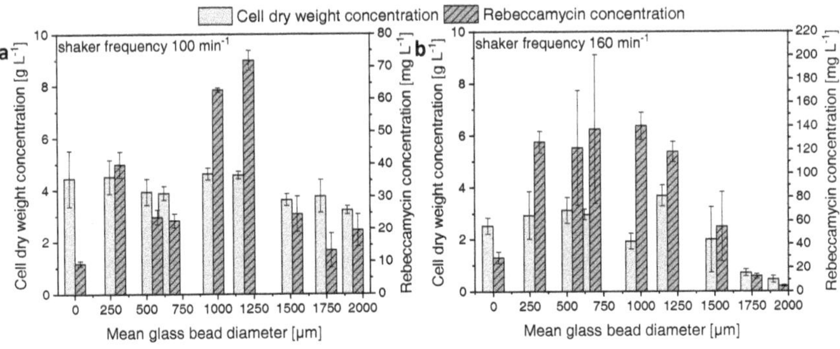

Figure 4.25: Cell dry weight and rebeccamycin concentrations a 10 day cultivation with addition of different glass bead diameters with a concentration of 100 g L^{-1} at different shaking frequencies of (a) 100 min^{-1} and (b) 160 min^{-1}.

For this investigation frequencies of 100 min^{-1} and 160 min^{-1} were chosen for cultivation to learn about effects of lower and higher frequencies than in a usual cultivation. With a shaking frequency of 100 min^{-1} (**Figure 4.25a**) the induced stress energy is reduced compared to conventional cultivations. With different glass bead diameters applied, a shift of the highest rebeccamycin concentration to a larger glass bead size was expected to compensate for the reduced shaking frequency. At 120 min^{-1} the mean glass bead size of 969 µm led to the maximal rebeccamycin titer at the given glass bead concentration (compare **Figure 4.14**). This shift can be observed in this approach. Addition of 969 µm glass beads resulted in a rebeccamycin concentration of 63 mg L^{-1} whereas the addition of 1183 µm glass beads led to 72 mg L^{-1} rebeccamycin. Significant differences in the CDWs for the different particle diameters are not observed.

With a shaking frequency of 160 min^{-1} (**Figure 4.25b**) the specific energy input is increased compared to regular cultivations. As a conclusion from the cultivation at 100 min^{-1}, a shift to a smaller particle size than 969 µm would be expected. However, due to large standard deviations at 658 and 540 µm mean diameter this shift can neither be clearly confirmed nor disproved. The maximal rebeccamycin titer (considering only the mean value) is observed at a glass bead diameter of 969 µm with 140 mg L^{-1} rebeccamycin. The total values of rebeccamycin concentrations differ significantly from the results in **Figure 4.25a**. However, they are not directly comparable since the cultivations are different biological approaches from different pre-cultures. What can be observed in this approach, are very low CDWs and rebeccamycin concentrations with addition of 1746 and 1932 µm glass beads (13 and 4 mg L^{-1} rebeccamycin, respectively). In previous cultivation these large particles did not result in such low rebeccamycin concentrations compared to the maximum and especially the CDWs were not as low as in this case with only 1.9 or 0.7 g L^{-1}. This indicates that the specific energy input for these particle sizes is too high resulting in cell destruction or even growth inhibition, which in turn reduces production. By additionally increasing the shaking frequency the specific energy input in this case was higher compared to a regular cultivation at 120 min^{-1}.

In general, the CDW in all approaches is rather similar for different specific energy inputs. However, in some cases a decreased CDW for the largest stress energy or stress frequency was observed. A cell destruction or growth inhibition could be a reason for this effect. The mostly similar CDW in the approaches supports the idea that the effects of glass beads are not biomass related, as already mentioned in **chapter 4.2.2**.

4.2.5 Effects of additional lecithin supplementation

The addition of lecithin to cultivations of actinomycetes isolated from soil was beneficial in previous studies regarding product formation (Brock 1956; Adelson et al. 1957). Due to the positive effects for similar microorganisms as *L. aerocolonigenes* lecithin supplemented cultivations were performed.

The addition of soy lecithin to the cultivation of *L. aerocolonigenes* was first conducted in different concentrations. This provides an overview of whether lecithin is beneficial and how much lecithin needs to be reasonably added for further investigations. The CDW and rebeccamycin concentrations for the addition of lecithin between 0 and 10 g L^{-1} are presented in **Figure 4.26**.

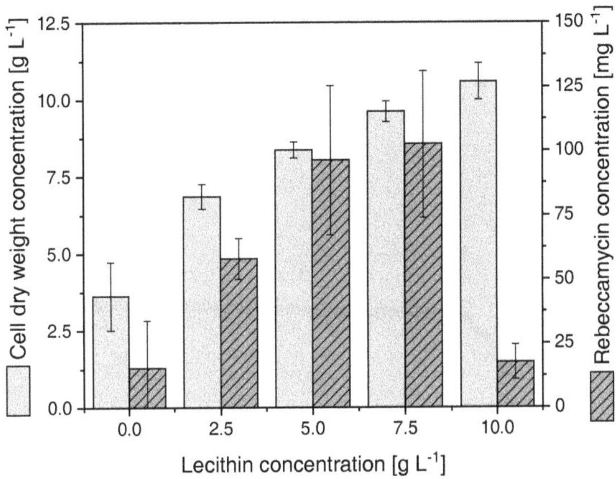

Figure 4.26: Cell dry weight and rebeccamycin concentrations of a 10 day cultivation with supplementation of soy lecithin in different concentrations.

The CDW is increasing with increasing lecithin concentration. The addition of 2.5 g L^{-1} lecithin almost doubles the biomass compared to an unsupplemented control, 10 g L^{-1} lecithin even results in a nearly 3-fold CDW. The rebeccamycin concentration, however, increases until 7.5 g L^{-1} lecithin (about 103 mg L^{-1} rebeccamycin) and is significantly lower for 10 g L^{-1} lecithin. The highest lecithin concentration results in a similar rebeccamycin concentration as the unsupplemented control (18 and 15 mg L^{-1}, respectively).

The increasing CDW suggests the metabolization of lecithin as an additional carbon source leading to enhanced growth. Larger amounts of biomass can produce larger amounts of rebeccamycin, explaining the increasing rebeccamycin concentration with increasing lecithin concentration. The decreased rebeccamycin concentration for 10 g L^{-1} lecithin might be explained with a prolonged exponential growth due to larger amounts of substrate. Since rebeccamycin is a secondary metabolite and is only produced after substrate deprivation, this could lead to a delayed start of product formation. To further investigate this hypothesis, the growth kinetics of a cultivation without and with 5 g L^{-1} lecithin were compared (**Figure 4.27**). 100 g L^{-1} of 969 µm glass beads were added to both apporaches.

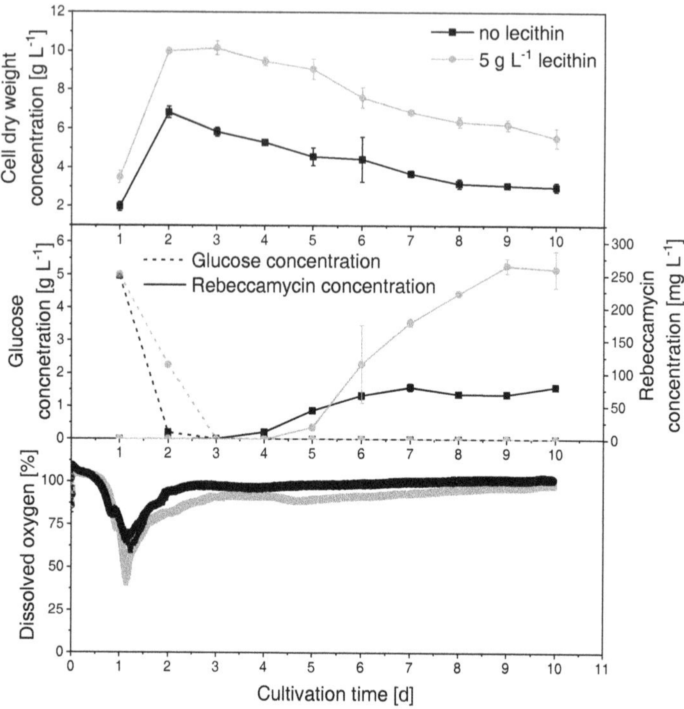

Figure 4.27: Cell dry weight, rebeccamycin and glucose concentrations as well as dissolved oxygen of a 10 day cultivation with 5 g L^{-1} and without lecithin. 100 g L^{-1} glass beads with a mean diameter of 969 µm were added in both approaches.

The CDWs for both approaches show a similar course, but the lecithin supplementation results in higher overall concentrations. The glucose consumption is slower for the lecithin supplemented approach. The glucose is fully depleted after 3 days, whereas without lecithin the glucose is almost completely depleted after 2 days. This affects the start of the rebeccamycin formation. Without lecithin the first amounts of rebeccamycin were measured on day 4 of cultivation while the addition of lecithin leads to a delayed rebeccamycin formation starting on day 5. However, in the approach with lecithin the rebeccamycin concentration then increases faster and to significantly higher concentrations (259 mg L^{-1} vs. 79 mg L^{-1} on day 10). The increased biomass growth with lecithin is also visible in the DO. The minimal value of the DO without lecithin is approximately 65 % whereas more biomass due to lecithin addition results in a minimal value of about 45 %. Moreover, the increase of the DO at the end of the exponential growth phase is slower for the lecithin addition (between day 2 and 3). These results support the hypothesis of a prolonged exponential growth of *L. aerocolonigenes* with lecithin supplementation.

Brock (1956) studied the influence of different oils and fatty acids on the filipin production in *Streptomyces filipinensis*. Among others, two different lecithins (vegetable and animal) were investigated. Both showed an increase in the product concentration compared to glucose as carbon source. However, the vegetable lecithin led to a twofold higher filipin concentration than the animal lecithin. The chosen lecithin for an increased rebeccamycin production in this thesis was made from soybean and might therefore potentially lead to higher product titers than animal lecithin. Lam et al. (1989) investigated different carbon sources for *L. aerocolonigenes*. Different carbohydrates were tested and starch resulted in the highest rebeccamycin concentration. However, since only defined carbohydrates were investigated no information about complex components like lecithin is given.

In **Figure 4.26** the addition of lecithin led to around 100 mg L^{-1} of rebeccamycin for 5 and 7.5 g L^{-1} of lecithin, which is a rather high product titer compared to literature (**Table 2.1**). Since lecithin acts as an additional carbon source, the combination of lecithin and glass beads could lead to even higher rebeccamycin titers as both methods employ different mode of actions for an increased production. Glass beads were already added in **Figure 4.27**, where the final rebeccamycin concentration with lecithin was even higher with 259 mg L^{-1}. However, since both experiments are different biological approaches started from different pre-cultures, a direct comparison is not totally reasonable. Therefore, a direct comparison with and without glass beads with different concentrations of lecithin was conducted for reliable results. The CDW and rebeccamycin concentrations of this approach are displayed in **Figure 4.28**.

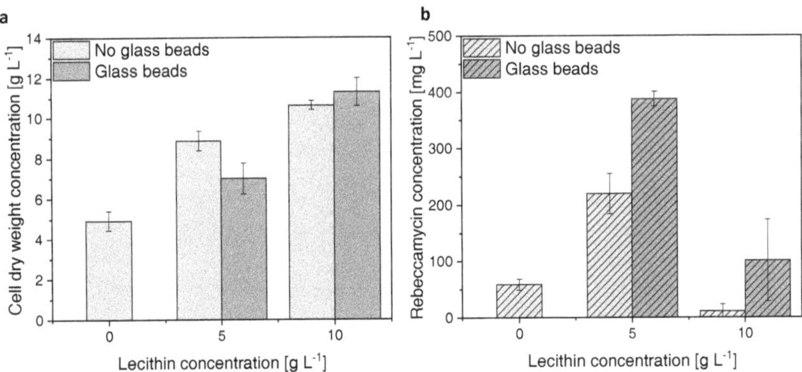

Figure 4.28: (a) Cell dry weight and (b) rebeccamycin concentrations of a 10 day cultivation with supplementation of soy lecithin in different concentrations and 100 g L^{-1} glass beads with a mean diameter of 969 µm.

The CDW, as already observed in **Figure 4.26**, is increasing with increasing lecithin concentration. For 5 g L^{-1} of lecithin the CDW of the cultivation with glass beads is lower than of the cultivation without glass beads, whereas for 10 g L^{-1} lecithin a similar level is observed. The rebeccamycin concentration for the cultivation with 5 g L^{-1} lecithin and no glass beads is around 220 mg L^{-1} and is therefore significantly higher than the cultivation without lecithin and no glass beads (59 mg L^{-1}). However, a combination of 5 g L^{-1} lecithin and 100 g L^{-1} glass beads further increases the rebeccamycin concentration to 388 mg L^{-1}. For 10 g L^{-1} lecithin the rebeccamycin concentration was significantly lower for both approaches.

In this approach a favorable substrate was combined with glass (macro) particle addition and resulted in an extremely high rebeccamycin concentration. The highest rebeccamycin concentration in shaking flasks achieved in literature was 183 mg L^{-1} (Nettleton, Jr. et al. 1985; Pommerehne et al. 2019). The given 388 mg L^{-1} rebeccamycin by addition of lecithin and glass particles is more than a twofold increase compared to this literature value.

The macro-morphological analysis of pellets from lecithin supplemented cultivations was inexplicit (data not shown). No conclusive correlation between any of the size or shape parameters and the increased production was observed. However, looking at the microscopic images in **Figure 4.29** some differences become apparent.

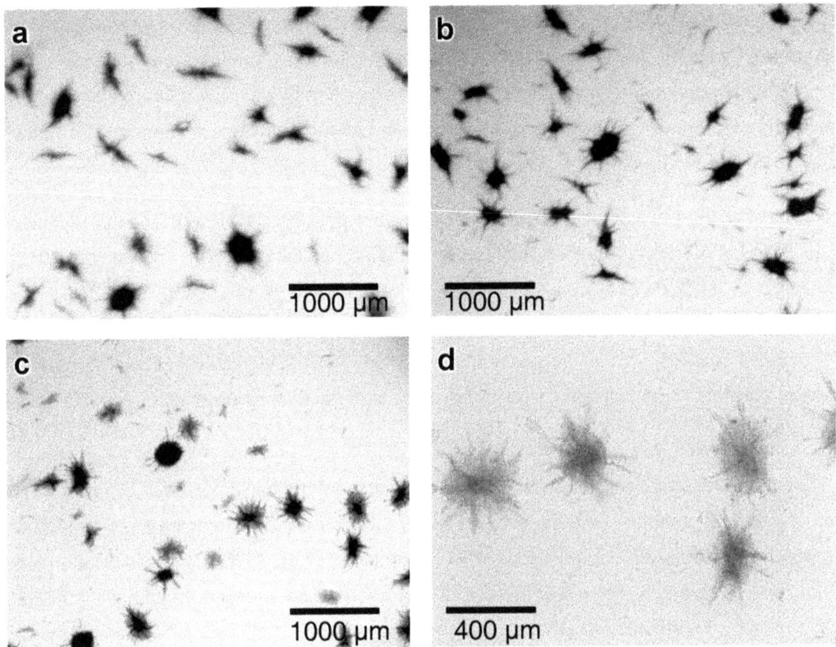

Figure 4.29: Microscopic images of *L. aerocolonigenes* (a) without lecithin and no glass beads, (b) with 5 g L^{-1} lecithin and no glass beads and (c, d) with 5 g L^{-1} lecithin and 100 g L^{-1} glass beads (Ø = 969 µm) in different magnifications.

Without glass beads, pellets from the unsupplemented cultivation and the cultivation with 5 g L^{-1} lecithin (**Figure 4.29a - b**) look similar. However, the combination of glass beads and lecithin results in pellets that, in some cases, appear less dense. This can be observed by a lighter color in the microscopic images (**Figure 4.29c**). At higher magnification (**Figure 4.29d**) a dense core and a less dense peripheral area of these pellets are visible. These pellets suggest changes in the micro-morphology or the inner pellet structure that might be connected to the significantly increased rebeccamycin formation. This was not yet further investigated but might be interesting for a more detailed morphological characterization.

4.3 Membrane aerated stirred bioreactor cultivations

For the production of larger amounts of rebeccamycin the transfer to a larger cultivation scale is necessary. However, information about bioreactor scale cultivations of *L. aerocolonigenes* in literature are scarce. Only Bush et al. (1987) described a successful cultivation in a 30 L stirred tank bioreactor cultivation. The conditions used in this cultivation were 27 °C, revolution of the stirrer of 375 min^{-1}, aeration rate of 60 L min^{-1} and 0.3 atm back pressure with the cultivation medium G134. This cultivation resulted in 663 mg L^{-1} of rebeccamycin, the highest rebeccamycin titer from a *L. aerocolonigenes* cultivation ever described in literature. However, no further details about reactor geometry, stirrer type and aeration type or challenges during the cultivation are given which makes it hard to learn from such results. Nonetheless, cultivations of *L. aerocolonigenes* in bioreactor scale are important for further development and are therefore part of this thesis.

The cultivation in a bioreactor offers many different possibilities of aeration and agitation, which should be carefully selected according to the application (e.g. Cerri and Badino (2012), Bliatsiou et al. (2019)). Bioreactor cultivations in this thesis were chosen to be performed using a bubble-free aeration technique. This decision was based on three reasons: 1) reduced foaming in the bioreactor, 2) simplified CFD-DEM-simulations for mechanical stress estimation (no bubbles need to be displayed), and 3) reduced input of mechanical stress.

The bubble-free aeration offers these advantages and can be implemented in different ways, e.g. by surface aeration, aeration through an external loop or bypass or aeration through an oxygen permeable membrane (Kopp 1989; Schneider et al. 1995; Lehmann et al. 1988) (compare **Figure 2.3**). For cultivations of *L. aerocolonigenes*, aeration through a membrane was chosen. Therefore, two different membrane aeration systems were developed and will be presented in the following chapters.

4.3.1 Aeration through a manually installed silicone tube

A thin-walled silicone tube (wall thickness $s = 0.2$ mm, inner diameter $d_i = 2.5$ mm, outer diameter $d_o = 2.9$ mm) was installed in the bioreactor by manually winding it around the reactor installations. Aeration was conducted with pure oxygen (aeration rate between 0.1 and 0.4 L min^{-1}) flowing through the silicone tube and diffusing into the cultivation medium. First test cultivations revealed the first challenge of this system: Membrane fouling. The closely spaced windings of the silicone tube allow the pellets to be trapped in between and to start growing on the membrane. This in turn can lead to problems regarding the diffusion of oxygen into the cultivation medium and leads to biomass loss in the cultivation broth. Therefore, membrane fouling should be avoided. Since for cultivations of *L. aerocolonigenes* the addition of glass beads was advantageous in regard to product formation in shake flasks, this was thought also to be useful in the case of preventing membrane fouling. A comparison of the membranes after 10 days of cultivation without glass beads and with 50 g (approximately 42 g L^{-1}) of ø = 969 µm glass beads is shown in **Figure 4.30**.

Figure 4.30: Membrane fouling after a 10 day cultivation (a) without and (b) with 50 g (approximately 42 g L^{-1}) of ø = 969 µm glass beads.

This glass bead concentration was chosen based on experiences from shake flask scale. In shake flasks, a glass bead concentration of about 100 g L^{-1} showed to be advantageous in cultivations of *L. aerocolonigenes*. However, since the glass beads induce mechanical stress and the mechanical stress in stirred bioreactors is generally larger than in shake flasks (Peter et al. 2006; Garcia-Ochoa et al. 2013), the glass bead concentration was reduced in order to reduce the stress events.

As it can be observed in the images of the membranes, the glass bead addition in this stirred bioreactor system has a positive effect on preventing membrane fouling. In the case of an unsupplemented cultivation (**Figure 4.30a**) excessive biomass growth on the membrane is visible, while after a cultivation with glass beads (**Figure 4.30b**) the biomass growth on the membrane is significantly reduced. The pellets are either not trapped between the tube windings because the glass beads are trapped even before inoculation and take up that space or the pellets are released again by the moving glass beads. These results indicate the benefit of using glass beads in this cultivation system. Therefore, glass beads were used for any further cultivations to reduce membrane fouling to a minimum.

Since there is only little information on bioreactor scale cultivations of *L. aerocolonigenes*, the initial cultivation should be used to provide information on parameters such as stirrer speed, aeration rate and length of the silicone membrane. Therefore, two bioreactors with silicone tubes of either 2.5 or 5 m length corresponding to a membrane exchange surface of 196 and 393 cm^2, respectively, were prepared. At the beginning of the cultivation, the stirrer speed was set to 500 min^{-1} and the aeration rate inside the silicone tube was 0.2 L min^{-1} of pure oxygen. Both parameters were adjusted during cultivation to avoid oxygen limitation. Results from these cultivations are displayed in **Figure 4.31**. The cultivations were conducted as single runs since they should only be used for an estimation of process parameters for further cultivations.

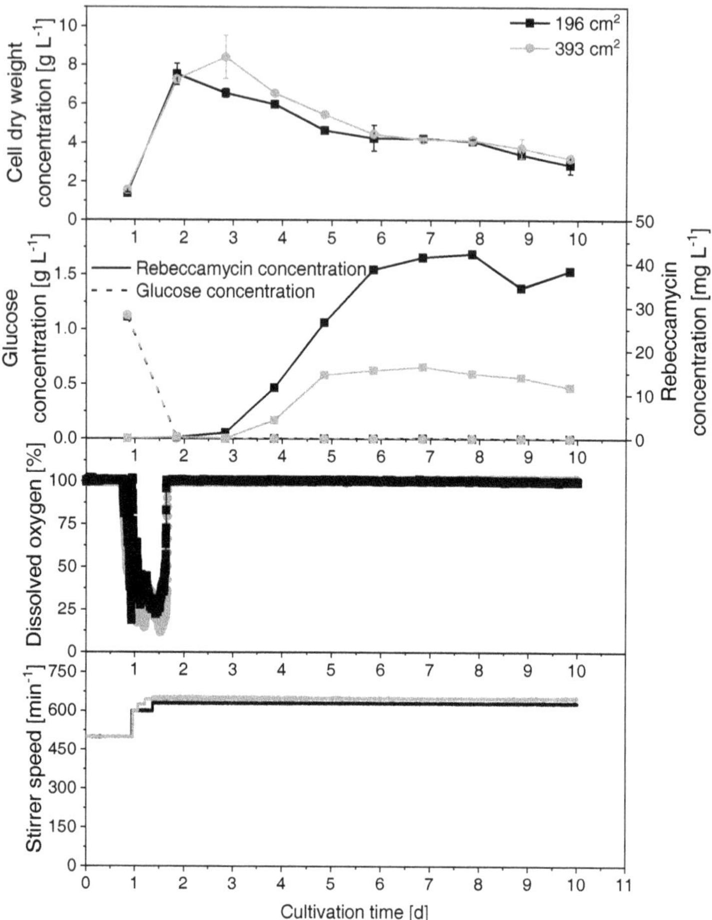

Figure 4.31: Cell dry weight, rebeccamycin and glucose concentrations as well as dissolved oxygen and stirrer speed of a 10 day cultivation in membrane aerated bioreactors with different membrane exchange surfaces of 196 cm^2 (black squares) and 393 cm^2 (grey circles). Both cultivations were glass bead supplemented (42 g L^{-1} of ø = 969 µm). The aeration rate (pure oxygen) was varied between 0.1 and 0.4 L min^{-1} and the stirrer speed was varied between 500 - 650 min^{-1}.

The CDWs are significantly increasing between day 1 and 2 for both approaches marking the exponential growth phase. From day 2 (membrane exchange surface of 196 cm^2) or day 3 (membrane exchange surface of 393 cm^2) the CDW decreases to around 3 g L^{-1}. The glucose concentration is nearly depleted after 2 days for both approaches and is therefore also marking the end of the exponential growth phase after 2 days of cultivation.

The rebeccamycin concentrations differ for both approaches. While for a membrane exchange surface of 393 cm^2 the maximal rebeccamycin concentration is 16 mg L^{-1}, the use of a membrane exchange surface of 196 cm^2 led to a maximal titer of 42 mg L^{-1}. To explain these product concentrations the DO during cultivation should be taken into account. Both approaches show a similar course of the DO. However, the minimal values are different. A membrane exchange surface of 196 cm^2 shows a decrease to a minimum of 22 % oxygen saturation whereas a membrane exchange surface of 393 cm^2 reached its minimum at 12 %. At first, this seems to be conflictive since the larger membrane exchange surface is expected to provide a higher oxygen input into the cultivation medium. However, another important factor might be the mixing through stirring. A larger membrane aeration surface also means a larger amount of silicone tube in the cultivation medium that is potentially disturbing the mixing (see also **Figure 3.1b** and **c**). A deteriorated mixing could also affect the rebeccamycin formation in *L. aerocolonigenes* and thereby lead to the lower rebeccamycin titers at a larger membrane exchange surface. Furthermore, Qi et al. (2003) described pressure differences along the length of membrane tubing. The pressure is highest at the aeration inlet and then decreases along the tubing. This also influences the oxygen transfer, that is lower at the exit of the tubing. This effect is larger, the longer the membrane tube. Hence, the membrane tubings used for aeration are often separated in several shorter segments to reduce this effect (Qi et al. 2003; Frahm et al. 2009).

The here presented cultivations were conducted to estimate reasonable values for cultivation process parameters. Therefore, the aeration rate, the stirrer speed and the membrane exchange surface should be considered. During the exponential growth phase (day 1 – 2 of cultivation) an adjustment of aeration rate and stirrer speed were necessary to avoid oxygen limitation. The aeration rate was set to 0.2 L min^{-1} at the start of the cultivation, was increased to 0.4 L min^{-1} at the beginning of the exponential growth phase (day 1) and was decreased to 0.1 L min^{-1} at the end of the exponential growth phase (day 2). The stirrer speed was additionally increased regularly from 500 to 650 min^{-1} between day 1 and 2 (see **Figure 4.31**). Since the mechanical stress is an important factor during cultivations of *L. aerocolonigenes*, as seen for the glass bead addition in shake flasks (see chapter **4.2**), this is something that will be further investigated in bioreactor scale. Hence, the stirrer speed will be set to a fixed value for a complete cultivation, since this facilitates the stress estimation. For the following cultivations, a stirrer speed of 650 min^{-1} was selected based on the results above, to guarantee a sufficient oxygen supply. The aeration rate was varied as described above, since there is an increased oxygen demand during the exponential growth phase and the membrane aeration does not impact the mechanical stress in the cultivation. Moreover, a membrane exchange surface of 196 cm^2 was chosen for further experiments as it seems to be advantageous in regard to rebeccamycin formation.

In the next cultivation, the influence of the glass bead concentration in the membrane aerated stirred bioreactor should be investigated. Thus, four different glass bead concentrations of 12.5, 25, 37.5 and 50 g L^{-1} (ø = 969 µm) were examined. The cultivation was conducted in duplicates. The results are presented in **Figure 4.32**.

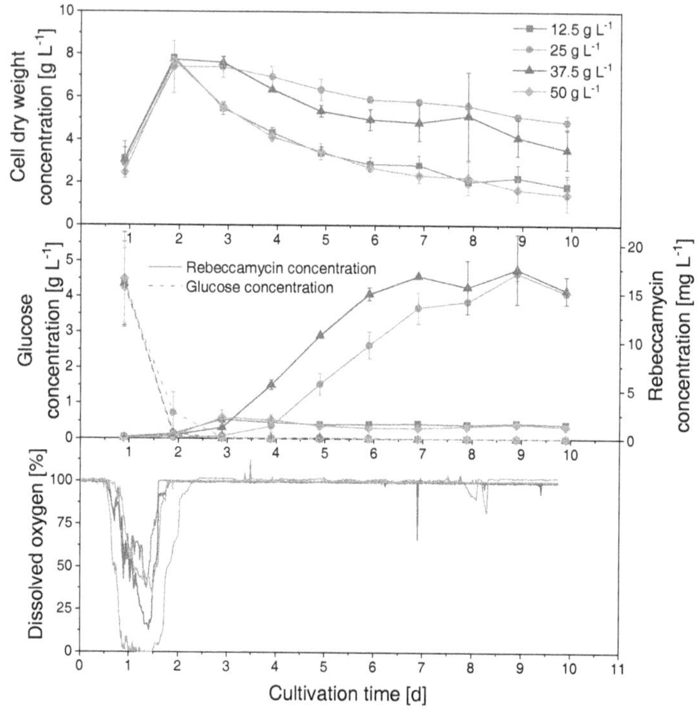

Figure 4.32: Cell dry weight, rebeccamycin and glucose concentrations as well as dissolved oxygen of a 10 day cultivation in membrane aerated bioreactors with different glass bead concentrations between 12.5 and 50 g L^{-1} (ø = 969 µm). The aeration rate (pure oxygen) was varied between 0.1 and 0.4 L min^{-1} with a stirrer speed of 650 min^{-1}.

The four different glass bead concentrations differently affect the cultivation of *L. aerocolonigenes*. Looking at the CDWs, all four approaches show a similar course until day 2, reaching a similar maximum of nearly 8 g L^{-1}. Thereafter, however, clear differences become apparent. The CDW for the glass bead concentrations of 25 and 37.5 g L^{-1} are slowly decreasing with final concentrations of 4.8 and 3.6 g L^{-1}, respectively, while the CDW for glass bead concentrations of 12.5 and 50 g L^{-1} are decreasing much faster with lower final concentrations of 1.8 and 1.4 g L^{-1}. The glucose consumption of all four approaches is similar and glucose is mostly depleted after day 2, though the glucose consumption with the addition of 25 g L^{-1} of glass beads is slightly lower with a complete depletion only after day 3.

The rebeccamycin concentrations show similar differences as the CDW. With a glass bead addition of 25 or 37.5 g L^{-1} the rebeccamycin titers reached a maximum of around 17 or 17.5 mg L^{-1} after 9 days of cultivation whereas glass bead concentrations of 12.5 or 50 g L^{-1} only led to a marginal rebeccamycin production. Moreover, the DO in the cultivation broth for 12.5 or 50 g L^{-1} glass bead addition only decreased to values of 34.6 and 33.7 %, respectively. With an addition of 37.5 g L^{-1} glass beads the DO was decreased to 13 % and with 25 g L^{-1} glass beads even to 0 % during the exponential growth phase.

Considering these results, glass bead concentrations between 25 and 37.5 g L^{-1} are the best choice in regard to product concentration under the given process conditions. Both glass bead concentrations lead to similar results. However, the rebeccamycin formation with 25 g L^{-1} glass beads seems to be slightly delayed compared to 35.7 g L^{-1} glass beads. This effect is likely caused by the oxygen limitation during the exponential growth phase. Liang et al. (2008) observed similar results during the natamycin production in *Streptomyces gilvosporeus*. A very low minimal DO of almost 0 %, caused by a low stirrer speed, led to a lower glucose consumption rate compared to two other approaches with a larger minimal DO. In this case, a lower product titer was observed additionally (Liang et al. 2008). The here presented rebeccamycin concentrations with 25 and 37.5 g L^{-1} glass beads are similar despite a different minimal DO. The lowest and highest applied glass bead concentration both led to very low rebeccamycin titers, proposing that either not enough or too much mechanical stress was induced (compare **chapter 4.2**). High glass bead concentrations can lead to the destruction of the pellets.

Another notable aspect is the overall lower rebeccamycin concentration with a maximum of 17.5 mg L^{-1} compared to the results from **Figure 4.31** with a maximum of 42 mg L^{-1} under similar process conditions. Both cultivations were started from different pre-cultures which may affect the resulting product titers. Differences of similar cultures from different pre-cultures were already observed in shake flasks (see **chapter 4.2**). However, another reason might be the gradual increase of the stirrer speed between day 1 and 2 in case of the first cultivation. A lower stirrer speed (500 min^{-1}) might be advantageous in the first period of the cultivation in regard to mechanical stress and the biomass development in this time period. An increase to a stirrer speed of 650 min^{-1} was necessary to avoid oxygen limitation. In different studies the occurrence of very low DO values led to lower or no production (Rosa et al. 2005; Liang et al. 2008) which is not desired. Although the start at a lower stirrer speed followed by an increase according to the DO seems to be beneficial in regard to rebeccamycin formation, this procedure was not used for further studies since the estimation of the overall induced mechanical stress by stirring and glass bead addition is impeded.

Therefore, the stirrer speed needs to be adjusted high enough to allow a sufficient oxygen supply but low enough to not destroy the microorganism during cultivation.

The results in this chapter show, that a cultivation of the filamentous microorganism *L. aerocolonigenes* in a membrane aerated stirred bioreactor is generally feasible and successfully provided rebeccamycin. However, there are still some drawbacks to the here presented membrane aeration system as described above. Hence, further improvements were made and are presented in the next chapters.

4.3.2 Characterization of a pressure-based membrane aeration system

Further development of the applied membrane aeration system was necessary to overcome the existing obstacles. Therefore, an improved membrane aeration system was designed and constructed by the project partner iPAT, TU Braunschweig, Germany (see also **Figure 3.2**). One of the major novelties was the change in geometry. In the membrane aeration system described above the silicone tube was manually installed in the bioreactor before each cultivation. This might lead to small differences in the geometry for every run and could already affect the cultivation itself. In this novel construction, the silicone membranes are in the same defined places for every cultivation. The amount of integrated membrane was chosen to be similar to the membrane aeration system described above. A further advantage of this defined geometry is the possibility of CFD-DEM-simulations for the estimation of the mechanical stress induced into the bioreactor analogous to the shake flask cultivations. Moreover, the windings of the silicone tube described above had only small spaces in between in which pellets could be trapped leading to cell growth on the membrane. This was circumvented by placing the tubes with a minimal distance of around 4 mm between each other which corresponds to a multiple of the maximum observed pellet diameter of *L. aerocolonigenes* (compare **Figure 4.17**). In this way the pellets do not get caught between the tubes and a cultivation without glass beads for the avoidance of membrane fouling might be possible. Furthermore, the application of a defined pressure to the silicone tubes was used with this new system. For the manually wound silicon tube described above, the aeration rate was increased during the exponential growth phase which was important for a sufficient oxygen supply. By increasing the aeration rate the pressure inside the tube was also increased. Increased pressure causes the silicone tube to expand, allowing an increased diffusion. Pressure can be applied to the improved membrane aeration system by a valve which is installed behind the oxygen outlet (compare **Figure 3.3** for a detailed description). In this way, the effect of different pressures can also be investigated.

The oxygen transfer through the silicone membrane into the cultivation medium can be characterized by the determination of the volumetric mass transfer coefficient (k_La). The k_La was determined via aeration with pure oxygen under different conditions. The pressure applied to the silicone tubes was varied between 0 and 0.75 bar. In this range, different aeration types were investigated. First, the aeration was conducted only through the silicone membrane, surplus gas was removed through the overpressure valve. In a second approach the surplus gas from the valve was redirected in the headspace of the bioreactor generating an additional surface aeration. For a third approach, not the surplus gas released from the valve was led to the bioreactor headspace, but an additional aeration source with a defined flow of 0.1 L min^{-1} oxygen was used for the surface aeration. The k_La values for these conditions are presented in **Figure 4.33**.

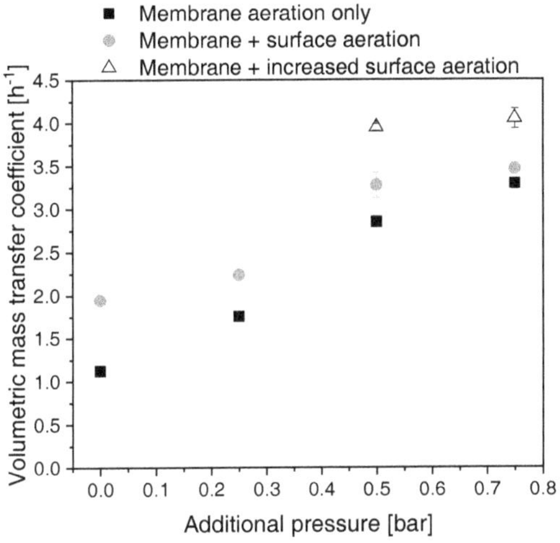

Figure 4.33: Volumetric mass transfer coefficient (k_La) at different pressures applied to the silicone tubes and a stirrer speed of 500 min^{-1}. The k_La of only membrane aeration (black squares), membrane and surface aeration (grey circles) as well as membrane aeration and an increased surface aeration (white triangles) was determined.

With increasing pressure applied to the silicone membrane the k_La increases for all approaches. The aeration only through the membrane results in the lowest k_La that rises linearly with increasing pressure applied to the membrane aeration system. Compared to these results the k_La could be increased with an additional surface aeration from the surplus gas released by the overpressure valve. Since less surplus gas is released by the valve at higher applied pressures, for 0.5 and 0.75 bar additional pressure on the silicone tube a further approach with a surface aeration from an independent source was measured. In this case the k_La was further increased reaching the overall maximum at 0.75 bar additional pressure with around 4 h^{-1}.

Compared to conventional cultivations in stirred bioreactors the total k_La values are very low. In a bioreactor with 2.2 L filling volume, aerated with 2.5 L min^{-1} air through an L-shaped sparger and agitated with two six-blade Rushton turbines a k_La of 56 h^{-1} was measured. However, a cultivation of *L. aerocolonigenes* in this system led to no rebeccamycin production (data not shown). Özbek and Gayik (2001) measured k_La values between 7.9 and 316.4 h^{-1} in a stirred bioreactor with a volume of 0.6 L with stirrer speeds of 100 - 500 min^{-1} in distilled water. Karimi et al. (2013) measured volumetric mass transfer coefficients in a stirred bioreactor with 1.7 L filling volume. With different configurations k_La values between 7.2 and 180 h^{-1} were measured. Furthermore, in the shake flasks used for cultivations (compare **chapters 4.1** and **4.2**) the k_La was measured to be around 52 h^{-1}. However, for membrane aeration systems the k_La is generally lower. Frahm et al. (2009) used a dynamic membrane aeration system with a silicone membrane and achieved volumetric mass transfer coefficients between 1 and 4 h^{-1}. Qi et al. (2003) measured a volumetric mass transfer coefficient of around 5 h^{-1} with a silicone membrane system used for aeration. They also tried to vary the exit pressure of the membrane tubing, however the k_La only increased slightly with increasing exit pressure (Qi et al. 2003).

Since the k_La with this membrane aeration system is rather low compared to e.g. shaking flasks, pure oxygen was used for aeration during cultivations. To visualize the differences between aeration with oxygen and air, the maximal oxygen transfer rate (OTR_{max}, equation (6)) was calculated for both and is presented in **Figure 4.34**.

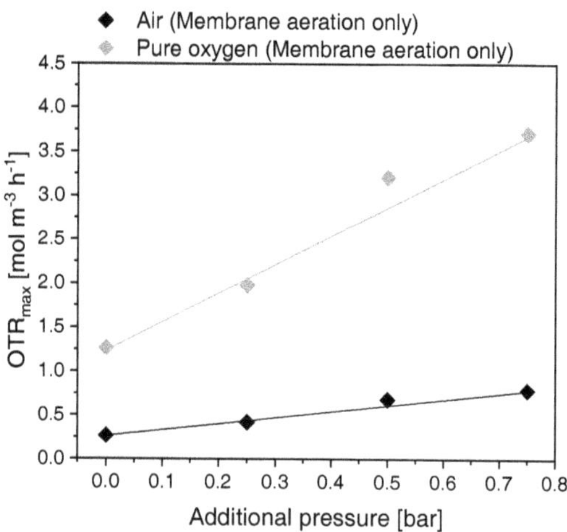

Figure 4.34: Maximal oxygen transfer rate (OTR_{max}) (calculated with equation (6)) at different pressures applied to the silicone tubes and a stirrer speed of 500 min^{-1}. The OTR_{max} with air or pure oxygen was determined. Membrane aeration alone was considered for comparison only.

The OTR_{max} for aeration with air or oxygen differ significantly from each other. For aeration with pure oxygen the OTR_{max} values are clearly higher. Especially at a larger pressure applied on the membrane system the difference between air and oxygen aeration is increasing. Hence, the aeration with pure oxygen contributes to an increased oxygen supply with the application this membrane aeration system.

The results presented in this chapter also justify the selection of silicone as a membrane material in this aeration system. The pressurized membrane aeration is only feasible with a non-porous material. When applying to much pressure to microporous membranes, the gas used for aeration will form bubbles on the membrane surface (Wagner and Lehmann 1988). Hence, a bubble-free aeration with pressure applied to the membrane system would not be possible, since the bubble-point is likely to be exceeded.

4.3.3 Cultivation in the pressure-based aeration system

To show the applicability of the pressure-based aeration system its use during a cultivation of *L. aerocolonigenes* is necessary. One idea of the newly developed membrane aeration module was the reduction of membrane fouling without glass bead addition by introducing larger spaces in between the silicone tubes. In this way the pellets of *L. aerocolonigenes* cannot get trapped between two silicone tubes. **Figure 4.35** shows the membrane aeration system after a 10 day cultivation without glass beads.

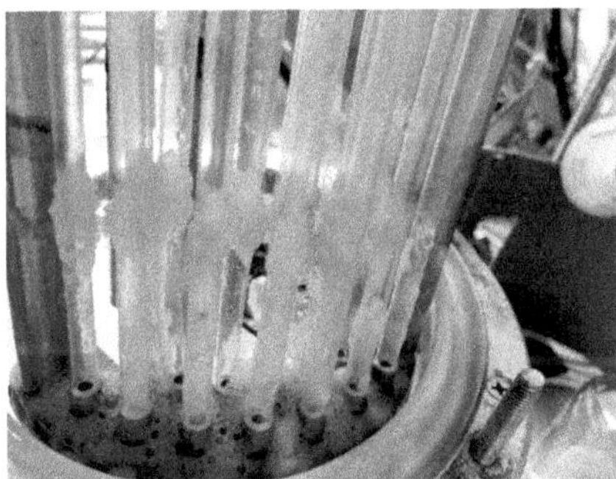

Figure 4.35: Membrane fouling after a 10 day cultivation with the pressurized membrane aeration system without glass bead addition at a stirrer speed of 650 min^{-1} and an additional pressure of 0.5 bar.

Some membrane fouling is visible in a certain height corresponding to the liquid level during cultivation. However, fouling between or on the silicone tubes is reduced to a minimum meaning that particle-free cultivations can be conducted in this system without much interference due to large areas of fouling as observed in **Figure 4.30a**. This allows reasonable cultivations with and without glass beads, in which the effects of glass beads can be compared to an unsupplemented cultivation.

Hence, two different cultivation approaches were conducted. One approach did not contain any glass beads and was cultivated at a stirrer speed of 650 min^{-1}. The second approach should be glass bead supplemented, but since an increased mechanical stress compared to the wound silicone tubes (compare **chapter 4.3.1**) due to a more massive installation was expected, the stirrer speed was reduced to 500 min^{-1} and only 25 g L^{-1} of 969 µm glass beads were added. For both approaches an additional pressure of 0.5 bar was applied. Although an increased k_La and OTR_{max} were observed for an additional pressure of 0.75 bar, this pressure was not feasible during cultivation. At this pressure leakages appeared after a certain time period causing bubbles in the cultivation medium. Hence, a lower pressure which was more stable during a long-term process was chosen. The results of the described approaches can be found in **Figure 4.36**.

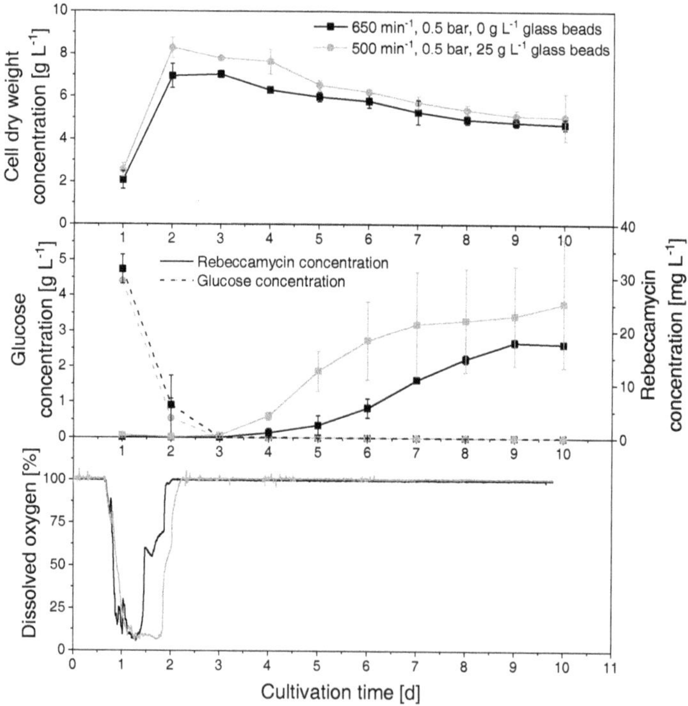

Figure 4.36: Cell dry weight, rebeccamycin and glucose concentrations as well as dissolved oxygen of a 10 day cultivation in a pressurized membrane aerated bioreactor with different conditions. Black squares: 650 min^{-1} stirrer speed, additional pressure on membrane of 0.5 bar, no glass beads; grey circles: 500 min^{-1} stirrer speed, 0.5 bar additional pressure on membrane and 25 g L^{-1} of 969 µm glass beads. Aeration was conducted using pure oxygen.

The course of the CDW is similar for both approaches, however, in the glass bead supplemented cultivation slightly larger CDWs were achieved. The maximal CDW was determined on day 2 with around 7.0 or 8.3 g L^{-1} without and with glass beads, respectively. The CDW was then decreasing until a final concentration of 4.7 or 5.0 g L^{-1}. Glucose was consumed similarly in both approaches. On day 3 no glucose was left in the cultivation broth.

The rebeccamycin production starts after day 3 of cultivation for both approaches. In the unsupplemented cultivation a maximal rebeccamycin concentration of around 18 mg L^{-1} was achieved. In the glass bead supplemented approach, the rebeccamycin concentration increases steeper in the beginning and on day 10 a mean of 25 mg L^{-1} rebeccamycin was provided. The DO decreases at a similar point in time for both approaches. However, the increase of the DO in the unsupplemented cultivation already occurs after ca. 1.5 days whereas in the cultivation with glass bead addition this increase only starts after around 2 days. The DO reaches a minimum of around 6 % oxygen saturation in both cases.

The maximal rebeccamycin titers obtained in these first cultivations with the pressurized membrane aeration system and without glass beads is similar to the maximal concentrations obtained with a manually wound silicone tube for aeration and glass bead concentrations of 25 or 37.5 g L^{-1} (compare **Figure 4.32**). Hence, a general applicability of this new system is given. The addition of 25 g L^{-1} glass beads to the cultivation led to a slightly increased rebeccamycin concentration compared to the approach without glass beads suggesting a similar effect as the addition of glass macroparticles already showed in shaking flask cultivations (compare **chapter 4.2**). However, the standard deviations of the rebeccamycin concentration with glass bead addition are comparatively high. For an overall assessment of the effects of glass macroparticle addition in bioreactors with this membrane aeration system a further investigation of glass bead concentration and size would be necessary.

To date, the cultivation of filamentous microorganisms in membrane aerated stirred bioreactors is only scarcely described in literature. Venkatadri and Irvine (1993) already used a silicone membrane aeration for the production of lignin peroxidase with the fungus *Phanerochaete chrysosporium*. However, in this case the growth of *P. chrysosporium* on the silicone membrane was desired in order to immobilize the microorganism, free growing biomass was not desired. The here presented cultivations of *L. aerocolonigenes* in a membrane aerated stirred bioreactor is the first application of such a system for the cultivation of a filamentous bacterium and the suspended cell growth in the cultivation medium. Due to the time-consuming development of this set-up further investigations such as the addition of microparticles or soy lecithin are reserved for future research.

4.4 Particle-based rebeccamycin recovery with adsorbent resins

Rebeccamycin recovery and purification has not yet been deeper investigated. Most procedures apply organic solvents for extraction and only use these extracts for analytical purposes (Pearce et al. 1988; Lam et al. 1989; Walisko et al. 2017). Before thinking of appropriate purification processes the process of rebeccamycin recovery should be considered in more detail, since the yields from extraction with organic solvent might not correspond to the potentially achievable yields.

The addition of adsorbent resins in the form of solid particles could be beneficial for this purpose. Different adsorbent resins are occasionally added to several filamentous microorganisms to bind valuable products. Mostly, an *in-situ* product recovery (ISPR) by adsorption of the product to the resins is desired. Additionally to this effect, often an increased product concentration was observed with adsorbent resin supplementation (Phillips et al. 2013).

4.4.1 XAD resin characterization

For the investigation of influences of adsorbent particles in this thesis three different particle types were selected, XAD 4, XAD 7 HP and XAD 16 N. The properties of these resins are presented in **Table 4.1**.

Table 4.1: Properties of the resins XAD 4, XAD 7 HP and XAD 16 N as given by the distributor (Sigma-Aldrich, Steinheim, Germany).

Property	XAD 4	XAD 7 HP	XAD 16 N
Material	Styrene-divenylbenzene	Acrylic ester	Styrene-divenylbenzene
Dipole moment	0.3	1.8	0.3
Density [g mL^{-1}]	1.02	1.05	1.02
Surface area [m^2 g^{-1}]	725	450	900
Porosity [mL mL^{-1}]	0.50	0.50	0.55
Pore volume [mL g^{-1}]	0.98	1.14	1.82
Pore diameter [Å]	50	90	100

XAD 4 and XAD 16 N are both consisting of the same material styrene-divenylbenzene (highly hydrophobic), whereas XAD 7 HP is made of acrylic ester (less hydrophobic). The acrylic ester has a higher dipole moment compared to the other resins, which might allow further interactions. Density and porosity of all resins are similar. The surface area is highest for XAD 16 N and lowest for XAD 7 HP, the adsorption capacity of XAD 16 N could therefore be highest. The pore diameter of XAD 4 is significantly lower than for the other resins. Hence, larger substances might only be adsorbed by XAD 7 HP or XAD 16 N.

A particle size range of the adsorbent resins was also given by the manufacturer; however, the particle size was additionally measured by laser diffraction (**Table 4.2**). To see the effects of stress on the adsorbent particles, the size was also measured after 24 h of stirring in distilled water.

Table 4.2: Particle size range of the adsorbent resins XAD 4, XAD 7 HP and XAD 16 N given by the distributor, measured median particle size and measured mean particle size after 24 h of stirring.

Particle size origin	XAD 4	XAD 7 HP	XAD 16 N
Particle size range (manufacturer) [µm]	490 - 690	560 - 710	560 – 710
x_{50} (measured) [µm]	540	579	635
x_{50} (measured after 24 h of stirring) [µm]	544	376	193

The measured median particle sizes arc located in the size range given by the manufacturer. XAD 4 particles are smallest, while XAD 16 N particles are largest. After 24 h of stirring XAD 4 remained a similar median size. However, XAD 7 HP and XAD 16 N decreased in median size, XAD 16 N even more than XAD 7 HP. XAD resins are composed of many small, agglomerated microspheres that form a single XAD particle (Puglisi et al. 2003). With mechanical stress induced on these XAD agglomerates, e.g., by stirring, they can be broken down into the individual microspheres. Frykman et al. (2006) addressed this topic and found significant changes in the particle size of adsorbent resins when stirred in a bioreactor for several days. Stirring can have a similar effect as shaking in shake flasks. Since the particles should be used in a cultivation which is constantly shaken for 10 days, this particle size might be important for the effects on *L. aerocolonigenes*.

The size of these adsorbent resins is significantly larger than the microparticles described in this thesis. Therefore, effects of incorporation or entanglement in between the pellets hyphae are unlikely to be observed. The adsorbent particles rather equal the larger glass and ceramic beads, classified as macroparticles in this thesis, in their size. The particles might therefore rather induce mechanical stress on the microorganism, but to a lesser extent due to the low density similar to water.

Before supplementing the resins in cultivations of *L. aerocolonigenes*, the general adsorption of rebeccamycin to the particles should be investigated. Moreover, the loss of rebeccamycin due to imperfect desorption from the resins should be considered. Results of a general adsorption test are illustrated in **Figure 4.37**. For this experiment, the adsorbent resins were added to the supernatant of a cultivation containing around 14 mg L^{-1} rebeccamycin. This value was used as 100 % and all other measured rebeccamycin concentrations were normalized according to this value. After adsorption particles and liquid phase were separated and rebeccamycin was eluted or extracted. The bars entitled "remained on particle" correspond to the calculated difference to 100 %.

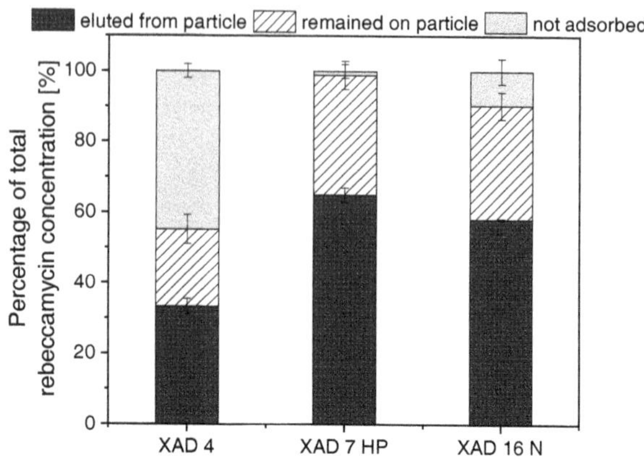

Figure 4.37: Adsorption and desorption of rebeccamycin on the adsorbent particles XAD 4, XAD 7 HP and XAD 16 N (Rebeccamycin concentrations were normalized according to a total concentration of 14 mg L^{-1} rebeccamycin. This value was used as 100 %).

As can be seen, a significant part (approximately 20 - 30 % depending on the resin) of the rebeccamycin is remaining on all three resin types and is not eluted. This means, that a loss of rebeccamycin is present. This loss of rebeccamycin should be compensated by an increased production when adsorbent particles are added to a cultivation to create a reasonable process. Furthermore, not all XAD resin types seem to adsorb rebeccamycin to the same extent. In case of XAD 4 nearly 45 % of rebeccamycin were not adsorbed and remained in the liquid phase. In case of XAD 16 N only around 10 % and in case of XAD 7 HP only approximately 1 % were not adsorbed to the resins. These results propose, that XAD 4 particles are not as suitable for the desired purpose as the other tested resins.

4.4.2 XAD resins in cultivations of *L. aerocolonigenes*

The next step was the addition of the resins in a cultivation of *L. aerocolonigenes*. The adsorbent particles were supplemented in a concentration of 50 g L^{-1} in each shaking flask since the addition of 5 % (in this case for 50 mL cultivation volume) of adsorbent resins was often used in literature (Phillips et al. 2013). The rebeccamycin concentrations determined at the end of the cultivation are displayed in **Figure 4.38**.

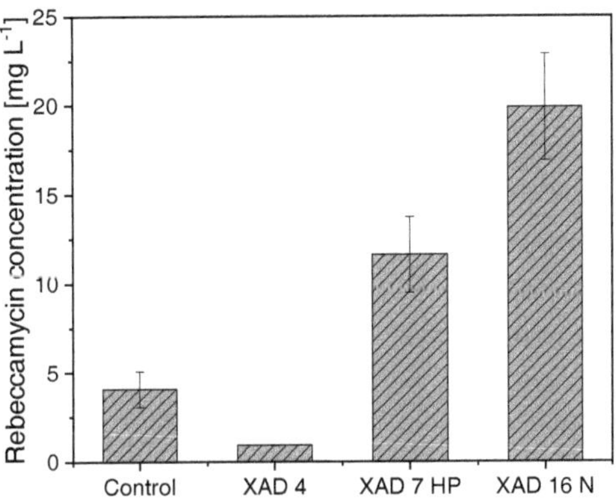

Figure 4.38: Rebeccamycin concentration of a 10 day shaking flask cultivation with XAD 4, XAD 7 HP and XAD 16 N resins in a concentration of 50 g L^{-1}.

Additionally, to the particle supplementations an unsupplemented control was prepared and cultivated under the same conditions. The rebeccamycin concentrations of the cultivations with the different resins are affected differently. XAD 4 supplementation shows a decreased rebeccamycin concentration of only around 1 mg L^{-1} compared to the unsupplemented control with 4 mg L^{-1}. XAD 7 HP and XAD 16 N supplementation, however, increased the rebeccamycin concentration compared to the control with 12 and 20 mg L^{-1}, respectively. These results imply a negative effect of XAD 4 addition in cultivations of *L. aerocolonigenes* in regard to product titer. Together with the lower rebeccamycin adsorption capacity of XAD 4, this particle type does not appear to be suitable for the intended purpose.

To find out about further effects of the resins during cultivation, the pellet morphology was investigated. The AESD and the roundness (equation (**11**)) are presented in **Figure 4.39**.

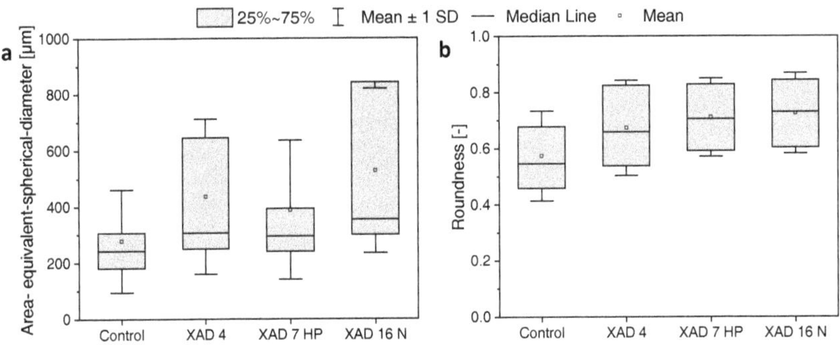

Figure 4.39: (a) Area-equivalent-spherical-diameter and (b) roundness (calculated with equation (11)) of pellets from a cultivation without or with the adsorbent particles XAD 4, XAD 7 HP or XAD 16 N. Pellets were attained from a cultivation with 50 g L^{-1} of particles.

The AESD (**Figure 4.39a**) of the pellets is slightly increased for adsorbent particle addition compared to pellets from an unsupplemented control. For XAD 4 and XAD 16 N the pellet diameter distribution is rather inhomogeneous. However, as already observed for the addition of glass macroparticles (see chapter **4.2.3**) the pellet size is not necessarily a meaningful indication for *L. aerocolonigenes*. Nevertheless, the roundness of pellets from a resin supplemented cultivation shows similar behavior as observed for the addition of glass macroparticles. The pellet roundness with addition of XAD particles is increased compared to untreated pellets. These results support the idea, that adsorbent particles rather effect the cultivation of *L. aerocolonigenes* by means of induced mechanical stress similar to glass macroparticles than similar to microparticles.

For a more detailed insight into the particle effects on the cultivation the CDW would be an interesting factor. However, due to the low density of the adsorbent particles and a partially similar size of the pellets (see **Figure 4.39a**), a separation of biomass and particles was hindered and the CDW was therefore not determined. Because of this aspect, the initial idea of an *in-situ* product removal would only be possible with a suitable separation technique of particles and pellets. Nevertheless, the process could still be facilitated using these adsorbent particles since rebeccamycin in the liquid phase could be adsorbed. Thus, only the solid phase is used for rebeccamycin recovery, reducing necessary volume for downstream processing.

Of course, the resins also have adsorbing effects during the cultivation. Since rebeccamycin is not present from the start of the cultivation other substances are prone to be adsorbed e.g., medium components that are essential for growth and production (González-Menéndez et al. 2014). Hence, this effect was investigated by pre-incubating the medium with the adsorbent particles followed by a cultivation in these media with the particles being removed before inoculation. The adsorption of substances was already visible before inoculation since the cultivation medium changed its color (**Figure 4.40**).

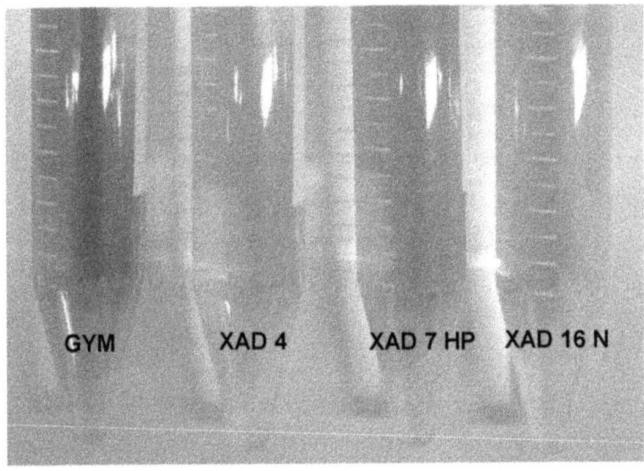

Figure 4.40: GYM medium after incubation with XAD 4, XAD 7 HP or XAD 16 N particles, untreated Gym medium (left) for comparison.

The changed color indicates the adsorption of medium colorants. Especially the incubation with XAD 4 and XAD 16 N led to a decoloration of the medium, whereas incubation with XAD 7 HP only brightened the medium. This could either be related to the material of the adsorbent resins or the dipole moment. These two properties are similar for XAD 4 and XAD 16 N and would therefore result in a similar adsorption. However, since an adsorption is visually recognized, the effects of this pre-incubation in cultivation of *L. aerocolonigenes* need to be investigated.

After pre-incubation the particles were removed and *L. aerocolonigenes* was cultivated in this medium (**Figure 4.41**). Since no resins were present during the cultivation, the determination of the CDW was facilitated and could therefore be carried out without obstruction.

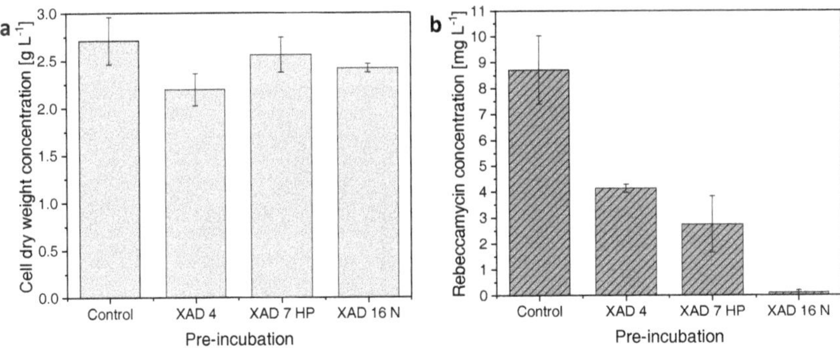

Figure 4.41: (a) Cell dry weight and (b) rebeccamycin concentrations of a 10 day cultivation pre-incubated with 50 g L^{-1} of XAD 4, XAD 7 HP or XAD 16 N.

A control without any kind of pre-incubation resulted in 2.7 g L^{-1} dry-biomass. The CDW in the pre-incubated approaches is located in a similar area. With XAD 4 pre-incubation 2.2 g L^{-1} CDW were achieved, pre-incubation with XAD 7 HP resulted in 2.6 g L^{-1} and with XAD 16 N in 2.4 g L^{-1}. In the control a rebeccamycin titer of around 9 mg L^{-1} was achieved. All approaches with pre-incubation and cultivation without glass beads resulted in lower rebeccamycin titer. Pre-incubation with XAD 4 led to approximately 4 mg L^{-1}, with XAD 7 HP to around 3 mg L^{-1} and with XAD 6 N only a marginal amount of rebeccamycin was produced.

Compared to the cultivation with adsorbent particles in the cultivation (**Figure 4.38**), this trend is opposite. All adsorbent resins reduce the final rebeccamycin concentration, suggesting that medium components important for the metabolism of *L. aerocolonigenes* are adsorbed. However, since the CDW at the end of the cultivation was similar for all approaches, the adsorption of medium components essential for growth seems to be unlikely. Considering this, rather substances which are essential for the production of rebeccamycin might be adsorbed. However, the search for the adsorbed substances would be extensive since a complex medium with a chemically not clearly defined composition is used. Nevertheless, the concentration of the amino acid tryptophan in pre-incubated medium was examined. Tryptophan is a precursor in the rebeccamycin biosynthesis of *L. aerocolonigenes* (Sánchez et al. 2006b). In **Figure 4.42** the tryptophan concentration in usual GYM medium, medium pre-incubated with the different adsorbent XAD resins and in the supernatant of an *L. aerocolonigenes* cultivation are displayed.

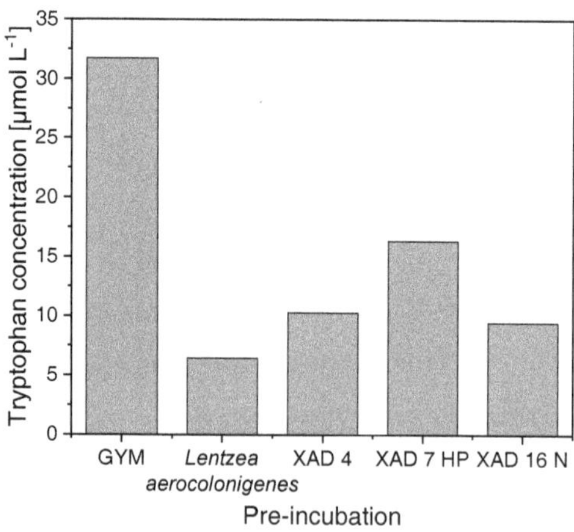

Figure 4.42: Tryptophan concentration in GYM medium, the supernatant of an *L. aerocolonigenes* cultivation.

In untreated GYM medium the tryptophan concentration is about 32 µmol L^{-1}. In all other approaches the tryptophan concentration was lower. The supernatant of a cultivation contained about 6 µmol L^{-1} tryptophan. This lower concentration indicates that tryptophan is consumed during cultivations of *L. aerocolonigenes*. The pre-incubation with XAD 4 and XAD 16 N resulted in a similar tryptophan concentration of around 10 µmol L^{-1} while the medium pre-incubated with XAD 7 HP contains 16 µmol L^{-1}. These results suggest that all three resins adsorb tryptophan. Lower tryptophan concentration in the medium correspond to a greater adsorption by the XAD resins, meaning the tryptophan adsorption is lowest for XAD 7 HP.

Furthermore, desorption effects might be important to consider. Substances adsorbed in the beginning of the cultivation may be exchanged by other substances during cultivation and can thereby be metabolized by the microorganism again. To further explain this idea, the findings of Kim et al. (2011) should be considered. Kim et al. (2011) also added different adsorbent resins to cultivations of an actinomycete and observed an increased product titer. However, in case of an avoidance of direct contact by separation of the resins from the microorganism using a material called miracloth, no increase in the product concentration was found. Kim et al. (2011) assumed that a direct transfer of the product from the mycelium to the resin occurred which was not possible under separation. This direct transfer might also be possible the other way around. The XAD resins adsorb essential substances, such as tryptophan, in the beginning of the cultivation and could directly transfer these substances to the mycelium during contact. In the case of a removal of the resins before the cultivation, these desorption effects cannot appear leading to a decreased product titer.

This direct transfer of substances from the resin to the microorganism would explain the differences in the rebeccamycin concentration with pre-incubation followed by resin removal and the permanent presence during the cultivation for XAD 7 HP and XAD 16 N. XAD 4 on the other hand, led to a decreased rebeccamycin titer compared to an unsupplemented control in both cases. The difference between the adsorbent resins might be related to the particle size of the XAD resins. As displayed in **Table 4.2**, the size of XAD 7 HP and XAD 16 N significantly decrease after 24 h of stirring while XAD 4 is stable in size. Due to the permanent shaking in the shake flasks during cultivation the resins XAD 7 HP and XAD 16 N will face a similar decease size. Smaller adsorbent particles offer a larger surface area, especially surface area which is not located inside of pores. Substances being adsorbed on the outer surface of a resin facilitates a potential direct transfer from the adsorbent resin to the microorganism.

In the next step, different concentrations of the resins in cultivation were investigated (**Figure 4.43**), since the so far applied concentration of 50 g L^{-1} was only a value from literature (Kim et al. 2011; Phillips et al. 2013). Due to the findings regarding XAD 4 in previous approaches, XAD 4 was considered to be not suitable for the desired purpose and is therefore not further investigated. Concentrations between 25 and 100 g L^{-1} were chosen since this area was also beneficial for the addition of similarly sized glass beads.

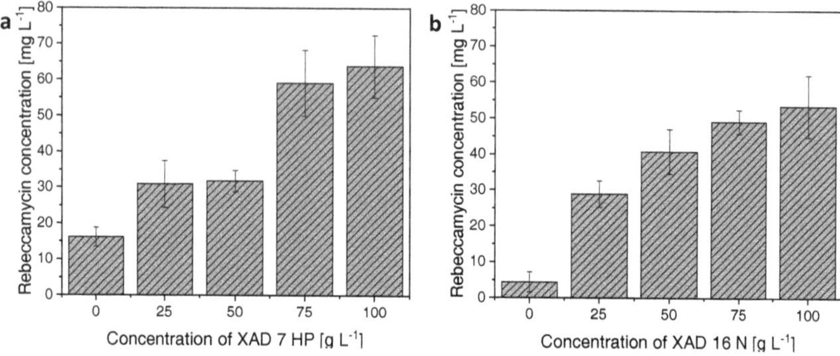

Figure 4.43: Rebeccamycin concentration depending on different particle concentrations of (a) XAD 7 HP and (b) XAD 16 N between 25 – 100 g L^{-1}.

The different concentrations of the adsorbent particles were tested in separate cultivation approaches inoculated from different pre-cultures. The total values of rebeccamycin concentrations are therefore not directly comparable, as can be observed in the different total values for the unsupplemented control in both approaches. However, with approaches started from the same pre-culture these deviations were not observed. For the addition of XAD 7 HP all investigated resin concentrations show an increased rebeccamycin concentration compared to the unsupplemented control. With increasing particle concentration, the rebeccamycin titer also increases. At 100 g L^{-1} of XAD 7 HP nearly 64 mg L^{-1} of rebeccamycin were produced which is a significant increase compared to 16 mg L^{-1} of rebeccamycin in the unsupplemented control. For the addition of XAD 16 N a similar trend is observed. The rebeccamycin concentration is increasing with increasing particle concentration. For the addition of 100 gL^{-1} of XAD 16 N a rebeccamycin concentration of 54 mg L^{-1} was achieved which is significantly higher than the unsupplemented control with around 4 mg L^{-1} rebeccamycin. The concentration of 100 g L^{-1} is most beneficial for cultivations with both types of particles. Moreover, this concentration, equaling 10 % in this case, fits the use of concentrations between 0.8 and 20 % described in literature (Phillips et al. 2013).

To find out whether part of the increase in rebeccamycin titer by adsorbent particles is caused by the induction of mechanical stress, not only the adsorption effects, additionally to the adsorbent particles glass particles were added (**Figure 4.44**). All particles were added in a concentration of 50 g L^{-1} in order to not exceed 100 g L^{-1} for a particle combination.

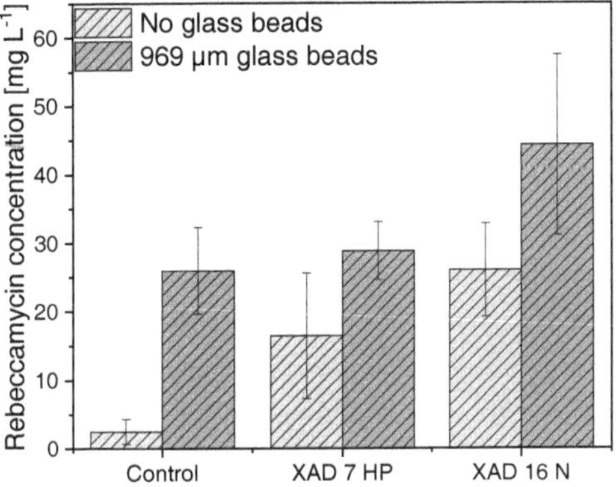

Figure 4.44: Rebeccamycin concentration of an unsupplemented control, XAD 7 HP or XAD 16 N supplementation (50 g L^{-1}) and with addition of 50 g L^{-1} of 969 µm glass beads to all approaches.

The addition of XAD 7 HP and XAD 16 N result in increased rebeccamycin concentrations (16 and 26 mg L^{-1}, respectively) compared to the unsupplemented control with nearly 3 mg L^{-1} as already observed in previous cultivations. However, a cultivation with 969 µm glass bead addition but no supplementation with adsorbent particles resulted in 26 mg L^{-1}, a similar titer as observed for XAD 16 N addition while XAD 7 HP addition generated a lower titer. The combination of XAD 7 HP and glass beads led to a rebeccamycin concentration of 29 mg L^{-1} which is only slightly increased compared to pure glass bead addition. Combining XAD 16 N and glass beads in a cultivation led to 44 mg L^{-1} rebeccamycin.

The results suggest that less mechanical stress is induced by the addition of the adsorbent particles than in case of glass bead addition. The adsorbent particle supplemented cultivations resulted in lower or similar rebeccamycin titers and adsorbing effects are additionally present. However, adsorbent particles inducing lower mechanical stress was already expected since the particles are smaller in size and have a lower density (around 1000 kg m^{-3} for adsorbent particles and 2500 kg m^{-3} for glass beads), both being important factors in the stress energy (equation (1)). However, a combination of both particle types is beneficial since adsorbing effects and effects caused by mechanical stress can be used together and lead to a further increase of the final product concentration.

5 Conclusions and future prospects

The aims of this thesis were the further investigation of micro- and macroparticle addition to cultivations of the filamentous *Lentzea aerocolonigenes* for an increased rebeccamycin production, the development and application of a bubble-free aerated bioreactor and a particle-based rebeccamycin recovery using adsorbent resins.

For microparticle supplemented cultivations of *L. aerocolonigenes* the entanglement of microparticles in the pellets and physicochemical surface effects of surface modified microparticles were to be investigated. Small glass beads were chosen as microparticles since the interaction of a glass bead surface with the pellets was assumed to be negligible. Glass microparticles with a median diameter of 7.9 µm in a concentration of 10 g L^{-1} increased the rebeccamycin concentration and were incorporated into the pellets. A viability staining showed an increased viability of pellets grown with glass microparticles compared to pellets from an unsupplemented approach. The surface modification of these glass microparticles with different carboxylic acids resulted in similar or increased rebeccamycin titers compared to the unsupplemented approach. The results support the importance of physicochemical surface effects of microparticles. However, none of these microparticles could exceed the rebeccamycin concentration achieved with untreated or NaOH treated glass microparticles. The surface modification with other substances might lead to even higher rebeccamycin titers. In all cases the pellet size was decreased with microparticle addition although differently modified microparticles were used, proposing that the macro-morphology is not the main driving force for the differences in the rebeccamycin titers. However, the surface modified microparticles led to different incorporation patterns in the pellets of *L. aerocolonigenes* (especially with HDA, MPA and PPA).

The effects of mechanical stress induced by macroparticles (diameters between 200 and 2100 µm) were investigated by a variation of particle diameter, particle concentration, particle density (ceramic vs. glass beads) and shaking frequency. In several approaches, signs of cell destruction and a decreased rebeccamycin concentration were observed, if the mechanical stress induced by the macroparticles was too high. However, if only low mechanical stress was induced (e.g. small particles or low concentrations) only a slight increase in the rebeccamycin concentration compared to an unsupplemented cultivation was observed. These results indicate the importance of the right amount of mechanical stress that can be induced on *L. aerocolonigenes* for the highest possible rebeccamycin concentration. The pellet morphology of *L. aerocolonigenes* without and with 100 g L^{-1} of 969 µm glass macroparticles was investigated in detail. The results from different cultivations conducted in different cultivation approaches showed that the area-equivalent-spherical-diameter and the circularity of pellets in some cases increased and in other cases decreased with the addition of glass beads. This led to the conclusion, that the macromorphology is not directly influenced by the glass bead addition. Therefore, the rebeccamycin production is assumed to not be directly linked to the macromorphology in this case. Similar behavior was already observed for other microorganisms (Belmar-Beiny and Thomas 1991; Yang et al. 1996). The investigation of the pellet micromorphology or the inner pellet structure might be interesting in this case, as nutrient and oxygen supply into the pellet core can be influenced by these factors.

In combination with the glass bead addition, the addition of soy lecithin to the cultivation medium was investigated in shaking flask scale. A lecithin concentration of 5 g L^{-1} combined with 100 g L^{-1} of 969 µm glass beads resulted in 388 mg L^{-1} rebeccamycin, the highest rebeccamycin concentration reported in literature in the shake flask scale (Pommerehne et al. 2019). Soy lecithin acted as an additional carbon source and was consumed by *L. aerocolonigenes* resulting in a higher CDW which in turn can produce more rebeccamycin.

The development of a bubble-free aerated stirred bioreactor was mainly intended for the reduction of foaming during the cultivation, a more precise adjustment of the mechanical stress and the facilitated stress estimation via CFD-DEM-simulations (conducted by the project partner iPAT, TU Braunschweig). A membrane aeration with a thin walled silicone tube was chosen for this purpose. In a first approach a silicone tube was installed in the bioreactor by winding it around the reactor installations. Glass bead addition during cultivation reduced fouling on the membrane and concentrations of 25 and 37.5 g L^{-1} glass beads resulted in about 17 mg L^{-1} rebeccamycin. For a second approach a membrane aeration system with a fixed geometry was used. For both approaches pure oxygen was used for aeration. In this case, pressure was applied to the membrane to increase the oxygen transfer in the cultivation broth. Fouling was reduced by an increased space between the silicone tubes, so that cultivations without glass bead addition were feasible. Without glass beads around 18 mg L^{-1} rebeccamycin were produced which is comparable to the rebeccamycin concentrations achieved in unsupplemented shake flask cultivations. In general, the rebeccamycin concentrations reached in these membrane aerated stirred bioreactor cultivations with glass bead addition are rather low compared to cultivations in shake flask scale. However, there are still several parameters which can be optimized, e.g., the stirrer geometry, the stirrer speed, the glass bead concentration or further parameters such as the inoculation volume or the addition of lecithin. Moreover, microparticle addition might also be beneficial in bioreactor cultivations as a production increase was already observed in shake flask scale. The highly promising addition of soy lecithin should also be considered in bioreactor scale. The cultivations presented in this thesis were the first application of a membrane aerated bioreactor for the cultivation of a filamentous bacterium. Submerged cultivations of filamentous microorganisms with free biomass were not performed before.

Adsorbent particles were to be applied for the rebeccamycin recovery and therefore a facilitated downstream processing. The adsorbent resins XAD 4, XAD 7 HP and XAD 16 N were added to cultivations of *L. aerocolonigenes* for this purpose. Whereas XAD 4 decreased the rebeccamycin concentration when added to the cultivation, XAD 7 HP and XAD 16 N increased the rebeccamycin concentration compared to an unsupplemented control. Moreover, XAD 4 could not completely adsorb rebeccamycin from an aqueous solution while in case of XAD 7 HP and XAD 16 N only marginal amounts of rebeccamycin are not adsorbed. The increased rebeccamycin titer with XAD 7 HP and XAD 16 N was suggested to be caused by the adsorption of the rebeccamycin precursor tryptophan from the cultivation medium and a direct transfer to the microorganism which used it for the rebeccamycin production. Furthermore, in one cultivation the adsorbent resins were combined with glass beads with a mean diameter of 969 µm which led to a further increase in the rebeccamycin titer. Both particles increase the rebeccamycin titer with different mechanisms. The glass beads induce mechanical stress on the microorganism while in case of the adsorbent resins adsorption effects are responsible. Both mechanisms do not seem to affect each other negatively and can therefore be used in combination.

One aspect that is apparent throughout all experiments in this thesis is the poor reproducibility of the rebeccamycin titer in cultivations of filamentous *L. aerocolonigenes*. Cultivations under the same conditions conducted at different times often result in different values for the rebeccamycin concentrations. These differences are mainly visible for cultivations started from different pre-cultures. Although the inoculation process was tried to be improved in this thesis, no satisfying process optimization was found. A further optimization of the inoculation process leading to an improved reproducibility would benefit the reliability of the rebeccamycin production with *L. aerocolonigenes* for future investigations.

In summary, the results presented in this thesis contribute to a better understanding of micro-, macro- or adsorbent particle addition to cultivations of *L. aerocolonigenes* and their effects on rebeccamycin production. Many of the future prospects described in this chapter will be investigated in a follow-up project in the DFG priority program 1934 *DiSPBiotech*.

6 List of abbreviations and symbols

Abbreviation	Meaning
ACN	Acetonitrile
APTES	3-Aminopropyltriethoxysilan
BA	Butanoic acid
CFD	Computational fluid dynamics
CLSM	Confocal laser scanning microscope
CPO	Chloroperoxidase
DA	Decanoic acid
DEM	Discrete element method
DIC	N,N'-Diisopropylcarbodiimid
EED	Down-pumping elephant ear impeller (elephant ear down-pumping)
FDA	Fluorescein diacetate
FLG	Green fluorescence channel
FLR	Red fluorescence channel
FSC	Forward scatter
GFP	Green fluorescent protein
GYM	Glucose-Yeast-Malt
HA	Hexanoic acid
HDA	Hexandioic acid
HPLC	High performance liquid chromatography
ISPR	*In-situ* product removal
Milli-Q	Highly purified water
MPA	3-Mercaptopropionic acid
MPEC	Microparticle-enhanced cultivation
PI	Propidium iodide
PP	Polypropylene
PPA	Phenylpropanoic acid
PTFE	Polytetrafluoroethylene
SSC	Sideward scatter
TFA	Trifluoric acid

Symbol	Meaning	Unit
AESD	Area-equivalent-spherical-diameter	µm
A_m	Membrane exchange surface	cm^2
c	Concentration	mol h^{-3}
c*	Saturation concentration	mol h^{-3}
CDW	Cell dry weight concentration	g L^{-1}
c_L	Oxygen concentration in the liquid phase	mol h^{-3}
d_{gm}	Grinding media diameter	m
d_i	Inner diameter	mm
d_o	Outer diameter	mm
DO	Dissolved oxygen	%
k_La	Volumetric mass transfer coefficient	h^{-1}
OTR	Oxygen transfer rate	mol m^{-3} h^{-1}
OUR	Oxygen uptake rate	mol m^{-3} h^{-1}
q_P	Biomass specific rebeccamycin productivity	mg g^{-1} d^{-1}
s	Wall thickness	mm
SE	Stress energy	N m
SF	Stress frequency	s^{-1}
SN	Stress number	-
STY	Space-time-yield	mg L^{-1} d^{-1}
t	Time	s
u_t	Stirrer tip speed	m s^{-1}
$\overline{x}$	Mean particle diameter	µm
x_{50}	Median particle diameter	µm
ρ_{gm}	Grinding media density	kg m^{-3}

7 References

Adelson LM, Schatz A, Trelawny GS (1957) Metabolism of lipids and lipid derivatives by a soil actinomycete. J Bacteriol 73:148–153

Anizon F, Golsteyn RM, Léonce S, Pfeiffer B, Prudhomme M (2009) A three-step synthesis from rebeccamycin of an efficient checkpoint kinase 1 inhibitor. Eur J Med Chem 44:2234–2238. https://doi.org/10.1016/j.ejmech.2008.05.023

Antecka A, Bizukojc M, Ledakowicz S (2016a) Modern morphological engineering techniques for improving productivity of filamentous fungi in submerged cultures. World J Microb Biot 32:193. https://doi.org/10.1007/s11274-016-2148-7

Antecka A, Blatkiewicz M, Bizukojć M, Ledakowicz S (2016b) Morphology engineering of basidiomycetes for improved laccase biosynthesis. Biotechnol Lett 38:667–672. https://doi.org/10.1007/s10529-015-2019-6

Barry DJ, Williams GA, Chan C (2015) Automated analysis of filamentous microbial morphology with AnaMorf. Biotechnol Progr 31:849–852. https://doi.org/10.1002/btpr.2087

Beinert S, Fragnière G, Schilde C, Kwade A (2018) Multiscale simulation of fine grinding and dispersing processes: Stressing probability, stressing energy and resultant breakage rate. Adv Powder Technol 29:573–583. https://doi.org/10.1016/j.apt.2017.11.034

Beinert S, Fragnière G, Schilde C, Kwade A (2015) Analysis and modelling of bead contacts in wet-operating stirred media and planetary ball mills with CFD–DEM simulations. Chem Eng Sci 134:648–662. https://doi.org/10.1016/j.ces.2015.05.063

Bellgardt K-H (1998) Process models for production of β-lactam antibiotics. In: Schügerl K (ed) Relation between morphology and process performances. Springer, Berlin, Heidelberg, pp 153–194. https://doi.org/10.1007/BFb0102282

Belmar-Beiny MT, Thomas CR (1991) Morphology and clavulanic acid production of *Streptomyces clavuligerus*: Effect of stirrer speed in batch fermentations. Biotechnol Bioeng 37:456–462. https://doi.org/10.1002/bit.260370507

Berovič M, Cimerman A (1979) Foaming in submerged citric acid fermentation on beet molasses. Eur J Appl Microbiol Biotechnol 7:313–319. https://doi.org/10.1007/BF00499845

Bizukojc M, Gonciarz J (2015) Influence of oxygen on lovastatin biosynthesis by *Aspergillus terreus* ATCC 20542 quantitatively studied on the level of individual pellets. Bioproc Biosyst Eng 38:1251–1266. https://doi.org/10.1007/s00449-015-1366-y

Bizukojc M, Ledakowicz S (2010) The morphological and physiological evolution of *Aspergillus terreus* mycelium in the submerged culture and its relation to the formation of secondary metabolites. World J Microb Biot 26:41–54. https://doi.org/10.1007/s11274-009-0140-1

Bliatsiou C, Schrinner K, Waldherr P, Tesche S, Böhm L, Kraume M, Krull R (2020) Rheological characteristics of filamentous cultivation broths and suitable model fluids. Biochem Eng J 163:107746. https://doi.org/10.1016/j.bej.2020.107746

Bliatsiou C, Malik A, Böhm L, Kraume M (2019) Influence of impeller geometry on hydromechanical stress in stirred liquid/liquid dispersions. Ind Eng Chem Res 58:2537–2550. https://doi.org/10.1021/acs.iecr.8b03654

Boruta T, Bizukojc M (2019) Application of aluminum oxide nanoparticles in *Aspergillus terreus* cultivations: Evaluating the effects on lovastatin production and fungal morphology. Biomed Res Int 2019:5832496. https://doi.org/10.1155/2019/5832496

Brock TD (1956) The effect of oils and fatty acids on the production of filipin. Appl Microbiol 4:131–133

Burkart MD (2003) Metabolic engineering - a genetic toolbox for small molecule organic synthesis. Org Biomol Chem 1:1–4. https://doi.org/10.1039/b210173d

Burstein HJ, Overmoyer B, Gelman R, Silverman P, Savoie J, Clarke K, Dumadag L, Younger J, Ivy P, Winer EP (2007) Rebeccamycin analog for refractory breast cancer. A randomized phase II trial of dosing schedules. Invest New Drug 25:161–164. https://doi.org/10.1007/s10637-006-9007-6

Bush JA, Long BH, Catino JJ, Bradner WT (1987) Production and biological activity of rebeccamycin, a novel antitumor agent. J Antibiot 40:668–678. https://doi.org/10.7164/antibiotics.40.668

Cairns TC, Feurstein C, Zheng X, Zheng P, Sun J, Meyer V (2019) A quantitative image analysis pipeline for the characterization of filamentous fungal morphologies as a tool to uncover targets for morphology engineering: a case study using *aplD* in *Aspergillus niger*. Biotechnol Biofuels 12:149. https://doi.org/10.1186/s13068-019-1473-0

Casini A, Chang F-Y, Eluere R, King AM, Young EM, Dudley QM, Karim A, Pratt K, Bristol C, Forget A, Ghodasara A, Warden-Rothman R, Gan R, Cristofaro A, Borujeni AE, Ryu M-H, Li J, Kwon Y-C, Wang H, Tatsis E, Rodriguez-Lopez C, O'Connor S, Medema MH, Fischbach MA, Jewett MC, Voigt C, Gordon DB (2018) A pressure test to make 10 molecules in 90 days. External evaluation of methods to engineer biology. J Am Chem Soc 140:4302–4316. https://doi.org/10.1021/jacs.7b13292

Cerri MO, Badino AC (2012) Shear conditions in clavulanic acid production by *Streptomyces clavuligerus* in stirred tank and airlift bioreactors. Bioproc Biosyst Eng 35:977–984. https://doi.org/10.1007/s00449-012-0682-8

Cronenberg CCH, Ottengraf SPP, Heuvel JC, Pottel F, Sziele D, Schgerl K, Bellgardt KH (1994) Influence of age and structure of *Pencillium chrysogenum* pellets on the internal concentration profiles. Bioprocess Eng 10:209–216. https://doi.org/10.1007/BF00369531

Dobson LF, O'Cleirigh CC, O'Shea DG (2008) The influence of morphology on geldanamycin production in submerged fermentations of *Streptomyces hygroscopicus var. geldanus*. Appl Microbiol Biot 79:859–866. https://doi.org/10.1007/s00253-008-1493-3

Doig SD, Pickering SCR, Lye GJ, Baganz F (2005) Modelling surface aeration rates in shaken microtitre plates using dimensionless groups. Chem Eng Sci 60:2741–2750. https://doi.org/10.1016/j.ces.2004.12.025

Dong M, Wang S, Xu F, Li Q, Li W (2018) Addition of aluminum oxide microparticles to *Trichoderma viride* My preculture enhances cellulase production and influences fungal morphology. Eng Life Sci 18:353–358. https://doi.org/10.1002/elsc.201700188

Driouch H, Hänsch R, Wucherpfennig T, Krull R, Wittmann C (2012) Improved enzyme production by bio-pellets of *Aspergillus niger*. Targeted morphology engineering using titanate microparticles. Biotechnol Bioeng 109:462–471. https://doi.org/10.1002/bit.23313

Driouch H, Roth A, Dersch P, Wittmann C (2011) Filamentous fungi in good shape: microparticles for tailor-made fungal morphology and enhanced enzyme production. Bioengineered bugs 2:100–104. https://doi.org/10.4161/bbug.2.2.13757

Driouch H, Sommer B, Wittmann C (2010a) Morphology engineering of *Aspergillus niger* for improved enzyme production. Biotechnol Bioeng 105:1058–1068. https://doi.org/10.1002/bit.22614

Driouch H, Roth A, Dersch P, Wittmann C (2010b) Optimized bioprocess for production of fructofuranosidase by recombinant *Aspergillus niger*. Appl Microbiol Biot 87:2011–2024. https://doi.org/10.1007/s00253-010-2661-9

Ehgartner D, Herwig C, Fricke J (2017) Morphological analysis of the filamentous fungus Penicillium chrysogenum using flow cytometry-the fast alternative to microscopic image analysis. Appl Microbiol Biot 101:7675–7688. https://doi.org/10.1007/s00253-017-8475-2

Ehgartner D, Fricke J, Schröder A, Herwig C (2016) At-line determining spore germination of Penicillium chrysogenum bioprocesses in complex media. Appl Microbiol Biot 100:8923–8930. https://doi.org/10.1007/s00253-016-7787-y

El-Enshasy H, Kleine J, Rinas U (2006) Agitation effects on morphology and protein productive fractions of filamentous and pelleted growth forms of recombinant *Aspergillus niger*. Process Biochem 41:2103–2112. https://doi.org/10.1016/j.procbio.2006.05.024

Elmayergi H, Scharer JM, Moo-Young M (1973) Effects of polymer additives on fermentation parameters in a culture of *A. niger*. Biotechnol Bioeng 15:845–859. https://doi.org/10.1002/bit.260150503

Etschmann MMW, Huth I, Walisko R, Schuster J, Krull R, Holtmann D, Wittmann C, Schrader J (2015) Improving 2-phenylethanol and 6-pentyl-α-pyrone production with fungi by microparticle-enhanced cultivation (MPEC). Yeast 32:145–157. https://doi.org/10.1002/yea.3022

Facompré M, Baldeyrou B, Bailly C, Anizon F, Marminon C, Prudhomme M, Colson P, Houssier C (2002) DNA targeting of two new antitumour rebeccamycin derivatives. Eur J Med Chem 37:925–932. https://doi.org/10.1016/S0223-5234(02)01423-X

Faul MM, Winneroski LL, Krumrich CA (1999) Synthesis of rebeccamycin and 11-dechlororebeccamycin. J Org Chem 64:2465–2470. https://doi.org/10.1021/jo982277b

Frahm B, Brod H, Langer U (2009) Improving bioreactor cultivation conditions for sensitive cell lines by dynamic membrane aeration. Cytotechnology 59:17–30. https://doi.org/10.1007/s10616-009-9189-9

Frykman S, Tsuruta H, Galazzo J, Licari P (2006) Characterization of product capture resin during microbial cultivations. J Ind Microbiol Biot 33:445–453. https://doi.org/10.1007/s10295-006-0088-1

Gallant M, Link JT, Danishefsky SJ (1993) A stereoselective synthesis of indole-β-N-glycosides. An application to the synthesis of rebeccamycin. J Org Chem 58:343–349. https://doi.org/10.1021/jo00054a015

Garcia-Ochoa F, Gomez E, Alcon A, Santos VE (2013) The effect of hydrodynamic stress on the growth of *Xanthomonas campestris* cultures in a stirred and sparged tank bioreactor. Bioproc Biosyst Eng 36:911–925. https://doi.org/10.1007/s00449-012-0825-y

Garcia-Ochoa F, Gomez E, Santos VE, Merchuk JC (2010) Oxygen uptake rate in microbial processes: An overview. Biochem Eng J 49:289–307. https://doi.org/10.1016/j.bej.2010.01.011

Garcia-Ochoa F, Gomez E (2009) Bioreactor scale-up and oxygen transfer rate in microbial processes. An overview. Biotechnol Adv 27:153–176. https://doi.org/10.1016/j.biotechadv.2008.10.006

Germec M, Yatmaz E, Karahalil E, Turhan İ (2017) Effect of different fermentation strategies on β-mannanase production in fed-batch bioreactor system. 3 Biotech 7:77. https://doi.org/10.1007/s13205-017-0694-9

Goel S, Wadler S, Hoffman A, Volterra F, Baker C, Nazario E, Ivy P, Silverman A, Mani S (2003) A phase II study of rebeccamycin analog NSC 655649 in patients with metastatic colorectal cancer. Invest New Drug 21:103–107. https://doi.org/10.1023/A:1022980613420

Gonciarz J, Kowalska A, Bizukojc M (2016) Application of microparticle-enhanced cultivation to increase the access of oxygen to *Aspergillus terreus* ATCC 20542 mycelium and intensify lovastatin biosynthesis in batch and continuous fed-batch stirred tank bioreactors. Biochem Eng J 109:178–188. https://doi.org/10.1016/j.bej.2016.01.017

Gonciarz J, Bizukojc M (2014) Adding talc microparticles to *Aspergillus terreus* ATCC 20542 preculture decreases fungal pellet size and improves lovastatin production. Eng Life Sci 14:190–200. https://doi.org/10.1002/elsc.201300055

González-Menéndez V, Asensio F, Moreno C, Pedro N de, Monteiro MC, La Cruz M de, Vicente F, Bills GF, Reyes F, Genilloud O, Tormo JR (2014) Assessing the effects of adsorptive polymeric resin additions on fungal secondary metabolite chemical diversity. Mycology 5:179–191. https://doi.org/10.1080/21501203.2014.942406

Henzler H-J (2000) Particle Stress in Bioreactors. In: Schügerl K, Kretzmer G, Henzler HJ, Kieran PM, MacLoughlin PE, Malone DM, Schumann W, Shamlou PA, Yim SS (ed) Influence of Stress on Cell Growth and Product Formation. Springer, Berlin, Heidelberg, pp 35–82. https://doi.org/10.1007/3-540-47865-5_2

Hille A, Neu TR, Hempel DC, Horn H (2009) Effective diffusivities and mass fluxes in fungal biopellets. Biotechnol Bioeng 103:1202–1213. https://doi.org/10.1002/bit.22351

Hille A, Neu TR, Hempel DC, Horn H (2005) Oxygen profiles and biomass distribution in biopellets of *Aspergillus niger*. Biotechnol Bioeng 92:614–623. https://doi.org/10.1002/bit.20628

Holtmann D, Vernen F, Müller JM, Kaden D, Risse JM, Friehs K, Dähne L, Stratmann A, Schrader J (2017) Effects of particle addition to *Streptomyces cultivations* to optimize the production of actinorhodin and streptavidin. Sustain Chem Pharm 5:67–71. https://doi.org/10.1016/j.scp.2016.09.001

Hotop S, Möller J, Niehoff J, Schügerl K (1993) Influence of the preculture conditions on the pellet size distribution of *Penicillium chrysogenum* cultivations. Process Biochem 28:99–104. https://doi.org/10.1016/0032-9592(93)80013-7

Hussain M, Vaishampayan U, Heilbrun LK, Jain V, LoRusso PM, Ivy P, Flaherty L (2003) A phase II study of rebeccamycin analog (NSC-655649) in metastatic renal cell cancer. Invest New Drug 21:465–471. https://doi.org/10.1023/A:1026259503954

Hyun C-G, Bililign T, Liao J, Thorson JS (2003) The biosynthesis of indolocarbazoles in a heterologous *E. coli* host. Chembiochem 4:114–117. https://doi.org/10.1002/cbic.200390004

Junker B (2007) Foam and its mitigation in fermentation systems. Biotechnol Prog 23:767–784. https://doi.org/10.1021/bp070032r

Junker BH (2004) Scale-up methodologies for *Escherichia coli* and yeast fermentation processes. J Biosci Bioeng 97:347–364. https://doi.org/10.1016/S1389-1723(04)70218-2

Kamilakis EG, Allen DG (1995) Cultivating filamentous microorganisms in a cyclone bioreactor. The influence of pumping on cell morphology. Process Biochem 30:353–360. https://doi.org/10.1016/0032-9592(95)87044-X

Kaneko T, Wong H, Utzig J, Schurig J, Doyle TW (1990) Water soluble derivatives of rebeccamycin. J Antibiot 43:125–127. https://doi.org/10.7164/antibiotics.43.125

Kaneko T, Wong H, Okamoto KT, Clardy J (1985) Two synthetic approaches to rebeccamycin. Tetrahedron Lett 26:4015–4018. https://doi.org/10.1016/S0040-4039(00)89281-3

Karahalil E, Coban HB, Turhan I (2019) A current approach to the control of filamentous fungal growth in media: microparticle enhanced cultivation technique. Crit Rev Biotechnol 39:192–201. https://doi.org/10.1080/07388551.2018.1531821

Karahalil E, Demirel F, Evcan E, Germeç M, Tari C, Turhan I (2017) Microparticle-enhanced polygalacturonase production by wild type *Aspergillus sojae*. 3 Biotech 7:361. https://doi.org/10.1007/s13205-017-1004-2

Karimi A, Golbabaei F, Mehrnia MR, Neghab M, Mohammad K, Nikpey A, Pourmand MR (2013) Oxygen mass transfer in a stirred tank bioreactor using different impeller configurations for environmental purposes. Iranian J Environ Health Sci Eng 10:6. https://doi.org/10.1186/1735-2746-10-6

Kaup B-A, Ehrich K, Pescheck M, Schrader J (2008) Microparticle-enhanced cultivation of filamentous microorganisms. Increased chloroperoxidase formation by *Caldariomyces fumago* as an example. Biotechnol Bioeng 99:491–498. https://doi.org/10.1002/bit.21713

Kieser T, Bibb MJ, Buttner MJ, Chater KF, Hopwood DA (2000) Practical *Streptomyces* genetics. John Innes Foundation, Norwich

Kim JJ, Lim SK, Lee MO, Lim SM, Lee B-M, Kim DH, Hyun J, Lee KS (2011) Method of extraction and yield-up of tricyclo compounds by adding a solid adsorbent resin as their carrier in fermentation medium. US Patent 20110183385 A1

Kockmann A, Hesselbach J, Schilde C, Kwade A, Garnweitner G (2016) Verbesserung von Kunstharzbeschichtungen durch Nanopartikel mit maßgeschneiderter Oberflächenmodifizierung. Chem Ing Tech 88:958–966. https://doi.org/10.1002/cite.201500171

Kopp T (1989) Blasenfreie Begasung. Acta Biotechnol 9:504–511

Kowalska A, Boruta T, Bizukojć M (2020) Performance of fungal microparticle-enhanced cultivations in stirred tank bioreactors depends on species and number of process stages. Biochem Eng J 161:107696. https://doi.org/10.1016/j.bej.2020.107696

Kowalska A, Boruta T, Bizukojc M (2019) Kinetic model to describe the morphological evolution of filamentous fungi during their early stages of growth in the standard submerged and microparticle-enhanced cultivations. Eng Life Sci 19:557–574. https://doi.org/10.1002/elsc.201900013

Kowalska A, Boruta T, Bizukojć M (2018) Morphological evolution of various fungal species in the presence and absence of aluminum oxide microparticles. Comparative and quantitative insights into microparticle-enhanced cultivation (MPEC). MicrobiologyOpen 6:e00603. https://doi.org/10.1002/mbo3.603

Kowalska A, Antecka A, Owczarz P, Bizukojć M (2017) Inulinolytic activity of broths of *Aspergillus niger* ATCC 204447 cultivated in shake flasks and stirred tank bioreactor. Eng Life Sci 17:1006–1020. https://doi.org/10.1002/elsc.201600247

Krull R, Wucherpfennig T, Esfandabadi ME, Walisko R, Melzer G, Hempel DC, Kampen I, Kwade A, Wittmann C (2013) Characterization and control of fungal morphology for improved production performance in biotechnology. J Biotechnol 163:112–123. https://doi.org/10.1016/j.jbiotec.2012.06.024

Kuhl M, Gläser L, Rebets Y, Rückert C, Sarkar N, Hartsch T, Kalinowski J, Luzhetskyy A, Wittmann C (2020) Microparticles globally reprogram Streptomyces albus towards accelerated morphogenesis, streamlined carbon core metabolism and enhanced production of the antituberculosis polyketide pamamycin. Biotechnology and bioengineering. https://doi.org/10.1002/bit.27537

Kwade A (2003) A stressing model for the description and optimization of grinding processes. Chem Eng Technol 26:199–205. https://doi.org/10.1002/ceat.200390029

Kwade A, Schwedes J (1997) Wet comminution in stirred media mills. KONA 15:91–102. https://doi.org/10.14356/kona.1997013

Labeda DP, Hatano K, Kroppenstedt RM, Tamura T (2001) Revival of the genus *Lentzea* and proposal for *Lechevalieria* gen. nov. Int J Syst Evol Microbiol 51:1045–1050. https://doi.org/10.1099/00207713-51-3-1045

Labeda DP (1986) Transfer of "*Nocardia aerocolonigenes*" (Shinobu and Kawato 1960) Pridham 1970 into the Genus *Saccharothrix* Labeda, Testa, Lechevalier, and Lechevalier 1984 as *Saccharothrix aerocolonigenes* sp. nov. Int J Syst Bacteriol 36:109–110. https://doi.org/10.1099/00207713-36-1-109

Lagmay JP, Krailo MD, Dang H, Kim A, Hawkins DS, Beaty O, Widemann BC, Zwerdling T, Bomgaars L, Langevin A-M, Grier HE, Weigel B, Blaney SM, Gorlick R, Janeway KA (2016) Outcome of patients with recurrent osteosarcoma enrolled in seven phase II trials through children's cancer group, pediatric oncology group, and children's oncology group. Learning from the past to move forward. J Clin Oncol 34:3031–3038. https://doi.org/10.1200/JCO.2015.65.5381

Lam KS, Mattei J, Forenza S (1989) Carbon catabolite regulation of rebeccamycin production in *Saccharothrix aerocolonigenes*. J Ind Microbiol 4:105–108. https://doi.org/10.1007/BF01569794

Langevin A-M, Bernstein M, Kuhn JG, Blaney SM, Ivy P, Sun J, Chen Z, Adamson PC (2008) A phase II trial of rebeccamycin analogue (NSC #655649) in children with solid tumors. A children's oncology group study. Pediatr Blood Cancer 50:577–580. https://doi.org/10.1002/pbc.21274

Lee JC, Park HR, Park DJ, Lee HB, Kim YB, Kim CJ (2003) Improved production of teicoplanin using adsorbent resin in fermentations. Lett Appl Microbiol 37:196–200. https://doi.org/10.1046/j.1472-765x.2003.01374.x

Lehmann J, Vorlop J, Büntemeyer H (1988) Bubble-free reactors and their development for continuous culture with cell recycle. In: Spier R, Griffiths JB (ed) Animal cell biotechnology, 3rd edn. Academic Press, pp 221–237

Leštan D, Leštan M, Chapelle JA, Lamar RT (1996) Biological potential of fungal inocula for bioaugmentation of contaminated soils. J Ind Microbiol 16:286–294. https://doi.org/10.1007/BF01570036

Liang J, Xu Z, Liu T, Lin J, Cen P (2008) Effects of cultivation conditions on the production of natamycin with *Streptomyces gilvosporeus* LK-196. Enzyme Microb Technol 42:145–150. https://doi.org/10.1016/j.enzmictec.2007.08.012

Li J, Ou D, Zheng L, Gan N, Song L (2011) Applicability of the fluorescein diacetate assay for metabolic activity measurement of *Microcystis aeruginosa* (Chroococcales, Cyanobacteria). Phycol Res 59:200–207. https://doi.org/10.1111/j.1440-1835.2011.00618.x

Lin P-J, Scholz A, Krull R (2010) Effect of volumetric power input by aeration and agitation on pellet morphology and product formation of *Aspergillus niger*. Biochem Eng J 49:213–220. https://doi.org/10.1016/j.bej.2009.12.016

Long B, Rose W, Vyas D, Matson J, Forenza S (2002) Discovery of antitumor indolocarbazoles. Rebeccamycin, NSC 655649, and fluoroindolocarbazoles. Curr Med Chem - Anti-Cancer Agents 2:255–266. https://doi.org/10.2174/1568011023354218

Luttmann R, Florek P, Preil W (1994) Silicone-tubing aerated bioreactors for somatic embryo production. Plant Cell Tissue Organ Cult 39:157–170. https://doi.org/10.1007/BF00033923

Maier U, Büchs J (2001) Characterisation of the gas–liquid mass transfer in shaking bioreactors. Biochem Eng J 7:99–106. https://doi.org/10.1016/S1369-703X(00)00107-8

Manteca A, Alvarez R, Salazar N, Yagüe P, Sanchez J (2008) Mycelium differentiation and antibiotic production in submerged cultures of *Streptomyces coelicolor*. Appl Environ Microbiol 74:3877–3886. https://doi.org/10.1128/AEM.02715-07

Marminon C, Pierré A, Pfeiffer B, Pérez V, Léonce S, Joubert A, Bailly C, Renard P, Hickman J, Prudhomme M (2003) Syntheses and antiproliferative activities of 7-azarebeccamycin analogues bearing one 7-azaindole moiety. J Med Chem 46:609–622. https://doi.org/10.1021/jm0210055

Morão A, Maia CI, Fonseca MMR, Vasconcelos JMT, Alves SS (1999) Effect of antifoam addition on gas-liquid mass transfer in stirred fermenters. Bioprocess Eng 20:165. https://doi.org/10.1007/s004490050576

Moreau P, Anizon F, Sancelme M, Prudhomme M, Sevère D, Riou JF, Goossens JF, Hénichart JP, Bailly C, Labourier E, Tazzi J, Fabbro D, Meyer T, Aubertin AM (1999) Synthesis, mode of action, and biological activities of rebeccamycin bromo derivatives. J Med Chem 42:1816–1822. https://doi.org/10.1021/jm980702n

Nettleton DE, Jr., Bush JA, Bradner WT, Doyle TW (1985) Process for producing rebeccamycin. US 4552842 A

Nielsen J (1996) Modelling the morphology of filamentous microorganisms. Trends Biotechnol 14:438–443. https://doi.org/10.1016/0167-7799(96)10055-X

Nieminen L, Webb S, Smith MCM, Hoskisson PA (2013) A flexible mathematical model platform for studying branching networks: experimentally validated using the model actinomycete, *Streptomyces coelicolor*. PloSONE 8:e54316. https://doi.org/10.1371/journal.pone.0054316

Niu K, Wu X-P, Hu X-L, Zou S-P, Hu Z-C, Liu Z-Q, Zheng Y-G (2020) Effects of methyl oleate and microparticle-enhanced cultivation on echinocandin B fermentation titer. Bioprocess Biosyst Eng. https://doi.org/10.1007/s00449-020-02389-3

Nock CJ, Brell JM, Bokar JA, Cooney MM, Cooper B, Gibbons J, Krishnamurthi S, Manda S, Savvides P, Remick SC, Ivy P, Dowlati A (2011) A phase I study of rebeccamycin analog in combination with oxaliplatin in patients with refractory solid tumors. Invest New Drug 29:126–130. https://doi.org/10.1007/s10637-009-9322-9

Nouioui I, Carro L, García-López M, Meier-Kolthoff JP, Woyke T, Kyrpides NC, Pukall R, Klenk H-P, Goodfellow M, Göker M (2018) Genome-based taxonomic classification of the phylum actinobacteria. Front Microbiol 9:2007. https://doi.org/10.3389/fmicb.2018.02007

Ochi K (1986) Occurrence of the stringent response in *Streptomyces sp.* and its significance for the initiation of morphological and physiological differentiation. J Gen Microbiol 132:2621–2631. https://doi.org/10.1099/00221287-132-9-2621

O'Cleirigh C, Casey JT, Walsh PK, O'Shea DG (2005) Morphological engineering of *Streptomyces hygroscopicus* var. *geldanus*. Regulation of pellet morphology through manipulation of broth viscosity. Appl Microbiol Biot 68:305–310. https://doi.org/10.1007/s00253-004-1883-0

Olmos E, Mehmood N, Haj Husein L, Goergen J-L, Fick M, Delaunay S (2013) Effects of bioreactor hydrodynamics on the physiology of *Streptomyces*. Bioprocess Biosyst Eng 36:259–272. https://doi.org/10.1007/s00449-012-0794-1

Onaka H, Ozaki T, Mori Y, Izawa M, Hayashi S, Asamizu S (2015) Mycolic acid-containing bacteria activate heterologous secondary metabolite expression in *Streptomyces lividans*. J Antibiot 68:594–597. https://doi.org/10.1038/ja.2015.31

Onaka H, Taniguchi S-i, Igarashi Y, Furumai T (2003a) Characterization of the biosynthetic gene cluster of rebeccamycin from *Lechevalieria aerocolonigenes* ATCC 39243. Biosci Biotechnol Biochem 67:127–138. https://doi.org/10.1271/bbb.67.127

Onaka H, Taniguchi S-i, Ikeda H, Igarashi Y, Furumai T (2003b) pTOYAMAcos, pTYM18, and pTYM19, actinomycete-*Escherichia coli* integrating vectors for heterologous gene expression. J Antibiot 56:950–956

Oncu S, Tari C, Unluturk S (2007) Effect of various process parameters on morphology, rheology, and polygalacturonase production by *Aspergillus sojae* in a batch bioreactor. Biotechnol Prog 23:836–845. https://doi.org/10.1021/bp070079c

Özbek B, Gayik S (2001) The studies on the oxygen mass transfer coefficient in a bioreactor. Process Biochem 36:729–741. https://doi.org/10.1016/S0032-9592(00)00272-7

Papagianni M (2014) Characterization of fungal morphology using digital image analysis techniques. J Microb Biochem Technol 06. https://doi.org/10.4172/1948-5948.1000142

Papagianni M (2004) Fungal morphology and metabolite production in submerged mycelial processes. Biotechnol Adv 22:189–259. https://doi.org/10.1016/j.biotechadv.2003.09.005

Park SW, Han SJ, Kim D-S, Sim SJ (2007) Improvement of epothilone B production by in situ removal of ammonium using cation exchange resin in *Sorangium cellulosum* culture. Biochem Eng J 37:328–331. https://doi.org/10.1016/j.bej.2007.05.013

Pearce CJ, Doyle TW, Forenza S, Lam KS, Schroeder DR (1988) The biosynthetic origins of rebeccamycin. J Nat Prod 50:937–940

Pekarsky A, Veiter L, Rajamanickam V, Herwig C, Grünwald-Gruber C, Altmann F, Spadiut O (2018) Production of a recombinant peroxidase in different glyco-engineered *Pichia pastoris* strains: a morphological and physiological comparison. Microb Cell Fact 17:183. https://doi.org/10.1186/s12934-018-1032-6

Peter CP, Suzuki Y, Büchs J (2006) Hydromechanical stress in shake flasks: correlation for the maximum local energy dissipation rate. Biotechnol Bioeng 93:1164–1176. https://doi.org/10.1002/bit.20827

Phillips DH (1966) Oxygen transfer into mycelial pellets. Biotechnol Bioeng 8:456–460. https://doi.org/10.1002/bit.260080311

Phillips T, Chase M, Wagner S, Renzi C, Powell M, DeAngelo J, Michels P (2013) Use of in situ solid-phase adsorption in microbial natural product fermentation development. J Ind Microbiol Biotechnol 40:411–425. https://doi.org/10.1007/s10295-013-1247-9

Pommerehne K, Walisko J, Ebersbach A, Krull R (2019) The antitumor antibiotic rebeccamycin-challenges and advanced approaches in production processes. Appl Microbiol Biot 103:3627–3636. https://doi.org/10.1007/s00253-019-09741-y

Pridham TG (1970) New names and new combinations in the order Actinomycetales Buchanan 1917. Technical Bulletin 1424. United States Department of Agriculture, US Government Printing Service, Washington D.C.

Puglisi E, Patterson CJ, Paton GI (2003) Non-exhaustive extraction techniques (NEETs) for bioavailability assessment of organic hydrophobic compounds in soils. Agronomie 23:755–756. https://doi.org/10.1051/agro:2003049

Qi HN, Goudar CT, Michaels JD, Henzler H-J, Jovanovic GN, Konstantinov KB (2003) Experimental and theoretical analysis of tubular membrane aeration for Mammalian cell bioreactors. Biotechnol Prog 19:1183–1189. https://doi.org/10.1021/bp025780p

Rajnisz A, Guśpiel A, Postek M, Ziemska J, Laskowska A, Rabczenko D, Solecka J (2016) Characterization and Optimization of Biosynthesis of Bioactive Secondary Metabolites Produced by *Streptomyces* sp. 8812. Pol J Microbiol 65:51–61. https://doi.org/10.5604/17331331.1197275

Ren X-D, Xu Y-J, Zeng X, Chen X-S, Tang L, Mao Z-G (2015) Microparticle-enhanced production of ε-poly- l -lysine in fed-batch fermentation. RSC Adv 5:82138–82143. https://doi.org/10.1039/C5RA14319E

Rischbieter E, Schumpe A, Wunder V (1996) Gas Solubilities in Aqueous Solutions of Organic Substances. J Chem Eng Data 41:809–812. https://doi.org/10.1021/je960039c

Rosa JC, Baptista Neto A, Hokka CO, Badino AC (2005) Influence of dissolved oxygen and shear conditions on clavulanic acid production by *Streptomyces clavuligerus*. Bioprocess Biosyst Eng 27:99–104. https://doi.org/10.1007/s00449-004-0386-9

Saberi A, Jalili H, Nikfarjam A, Koohsorkhi J, Jarmoshti J, Bizukojc M (2020) Monitoring of *Aspergillus terreus* morphology for the lovastatin production in submerge culture by impedimetry. Biochem Eng J 159:107615. https://doi.org/10.1016/j.bej.2020.107615

Sánchez C, Méndez C, Salas JA (2006a) Engineering biosynthetic pathways to generate antitumor indolocarbazole derivatives. J Ind Microbiol Biotechnol 33:560–568. https://doi.org/10.1007/s10295-006-0092-5

Sánchez C, Méndez C, Salas JA (2006b) Indolocarbazole natural products. Occurrence, biosynthesis, and biological activity. Nat Prod Rep 23:1007–1045. https://doi.org/10.1039/b601930g

Sánchez C, Butovich IA, Braña AF, Rohr J, Méndez C, Salas JA (2002) The biosynthetic gene cluster for the antitumor rebeccamycin. Characterization and generation of indolocarbazole derivatives. Chem Biol 9:519–531

Saruyama N, Sakakura Y, Asano T, Nishiuchi T, Sasamoto H, Kodama H (2013) Quantification of metabolic activity of cultured plant cells by vital staining with fluorescein diacetate. Anal Biochem 441:58–62. https://doi.org/10.1016/j.ab.2013.06.005

Schmideder S, Barthel L, Friedrich T, Thalhammer M, Kovačević T, Niessen L, Meyer V, Briesen H (2019a) An X-ray microtomography-based method for detailed analysis of the three-dimensional morphology of fungal pellets. Biotechnol Bioeng 116:1355–1365. https://doi.org/10.1002/bit.26956

Schmideder S, Barthel L, Müller H, Meyer V, Briesen H (2019b) From three-dimensional morphology to effective diffusivity in filamentous fungal pellets. Biotechnol Bioeng 116:3360–3371. https://doi.org/10.1002/bit.27166

Schneider M, Reymond F, Marison IW, Stockar U von (1995) Bubble-free oxygenation by means of hydrophobic porous membranes. Enzyme Microb Technol 17:839–847. https://doi.org/10.1016/0141-0229(94)00113-6

Schrader M, Pommerehne K, Wolf S, Finke B, Schilde C, Kampen I, Lichtenegger T, Krull R, Kwade A (2019) Design of a CFD-DEM-based method for mechanical stress calculation and its application to glass bead-enhanced cultivations of filamentous *Lentzea aerocolonigenes*. Biochem Eng J 148:116–130. https://doi.org/10.1016/j.bej.2019.04.014

Schrinner K, Veiter L, Schmideder S, Doppler P, Schrader M, Münch N, Althof K, Kwade A, Briesen H, Herwig C, Krull R (2020) Morphological and physiological characterization of filamentous *Lentzea aerocolonigenes*: Comparison of biopellets by microscopy and flow cytometry. PloS one 15:e0234125. https://doi.org/10.1371/journal.pone.0234125

Schwandt A, Mekhail T, Halmos B, O'Brien T, Ma PC, Fu P, Ivy P, Dowlati A (2012) Phase-II trial of rebeccamycin analog, a dual topoisomerase-I and -II inhibitor, in relapsed "sensitive" small cell lung cancer. J Thorac Oncol 7:751–754. https://doi.org/10.1097/JTO.0b013e31824abca2

Shinobu R, Kawato M (1960) On *Streptomyces aerocolonigenes nov. sp.*, forming the secondary colonies on the aerial mycelia. Bot Mag 73:212–216

Singh MP, Leighton MM, Barbieri LR, Roll DM, Urbance SE, Hoshan L, McDonald LA (2010) Fermentative production of self-toxic fungal secondary metabolites. J Ind Microbiol Biotechnol 37:335–340. https://doi.org/10.1007/s10295-009-0678-9

Sohoni SV, Bapat PM, Lantz AE (2012) Robust, small-scale cultivation platform for *Streptomyces coelicolor*. Microb Cell Fact 11:9. https://doi.org/10.1186/1475-2859-11-9

Stocks SM (2004) Mechanism and use of the commercially available viability stain, *Bac*Light. Cytometry A 61:189–195. https://doi.org/10.1002/cyto.a.20069

Tamura S, Park Y, Toriyama M, Okabe M (1997) Change of mycelial morphology in tylosin production by batch culture of *Streptomyces fradiae* under various shear conditions. J Ferment Bioeng 83:523–528. https://doi.org/10.1016/S0922-338X(97)81131-2

Tao T-L, Cui F-J, Chen X-X, Sun W-J, Huang D-M, Zhang J, Yang Y, Di Wu, Liu W-M (2018) Improved mycelia and polysaccharide production of *Grifola frondosa* by controlling morphology with microparticle Talc. Microb Cell Fact 17:1. https://doi.org/10.1186/s12934-017-0850-2

Taticek RA, Moo-Young M, Legge RL (1991) The scale-up of plant cell culture: Engineering considerations. Plant Cell Tissue Organ Cult 24:139–158. https://doi.org/10.1007/BF00039742

Tesche S, Rösemeier-Scheumann R, Lohr J, Hanke R, Büchs J, Krull R (2019) Salt-enhanced cultivation as a morphology engineering tool for filamentous actinomycetes: Increased production of labyrinthopeptin A1 in *Actinomadura namibiensis*. Eng Life Sci 19:781–794. https://doi.org/10.1002/elsc.201900036

Tough AJ, Prosser JI (1996) Experimental verification of a mathematical model for pelleted growth of *Streptomyces coelicolor* A3(2) in submerged batch culture. Microbiology 142 (Pt 3):639–648. https://doi.org/10.1099/13500872-142-3-639

Tsueng G, Lam KS (2007) Stabilization effect of resin on the production of potent proteasome inhibitor NPI-0052 during submerged fermentation of *Salinispora tropica*. J Antibiot 60:469–472. https://doi.org/10.1038/ja.2007.61

Vardar-Sukan F (1992) Foaming and its Control in Bioprocesses. In: Vardar-Sukan F, Sukan ŞS (ed) Recent Advances in Biotechnology. Kluwer Academic Publishers, Dordrecht, Boston, pp 113–146. https://doi.org/10.1007/978-94-011-2468-3_6

Vecht-Lifshitz SE, Sasson Y, Braun S (1992) Nikkomycin production in pellets of *Streptomyces tendae*. J Appl Bacteriol 72:195–200. https://doi.org/10.1111/j.1365-2672.1992.tb01823.x

Vecht-Lifshitz SE, Magdassi S, Braun S (1990) Pellet formation and cellular aggregation in *Streptomyces tendae*. Biotechnol Bioeng 35:890–896. https://doi.org/10.1002/bit.260350906

Veiter L, Herwig C (2019) The filamentous fungus *Penicillium chrysogenum* analysed via flow cytometry - a fast and statistically sound insight into morphology and viability. Appl Microbiol Biot 103:6725–6735. https://doi.org/10.1007/s00253-019-09943-4

Venkatadri R, Irvine RL (1993) Cultivation of *Phanerochaete chrysosporium* and production of lignin peroxidase in novel biofilm reactor systems: Hollow fiber reactor and silicone membrane reactor. Water Res 27:591–596. https://doi.org/10.1016/0043-1354(93)90168-H

Wagner R, Lehmann J (1988) The growth and productivity of recombinant animal cells in a bubble-free aeration system. Trends Biotechnol 6:101–104

Walisko J (2017) Morphologiebeeinflussung von *Lechevalieria aerocolonigenes* und heterologe Produktion von Rebeccamycin. ibvt-Schriftenreihe (Krull R (ed.)), Vol. 78, Cuvillier Verlag, Göttingen. Dissertation, Technische Universtiät Braunschweig.

Walisko J, Vernen F, Pommerehne K, Richter G, Terfehr J, Kaden D, Dähne L, Holtmann D, Krull R (2017) Particle-based production of antibiotic rebeccamycin with *Lechevalieria aerocolonigenes*. Process Biochem 53:1–9. https://doi.org/10.1016/j.procbio.2016.11.017

Walisko R, Moench-Tegeder J, Blotenberg J, Wucherpfennig T, Krull R (2015) The taming of the shrew - Controlling the morphology of filamentous eukaryotic and prokaryotic microorganisms. Adv Biochem Eng Biotechnol 149:1–27. https://doi.org/10.1007/10_2015_322

Walisko R, Krull R, Schrader J, Wittmann C (2012) Microparticle based morphology engineering of filamentous microorganisms for industrial bio-production. Biotechnol Lett 34:1975–1982. https://doi.org/10.1007/s10529-012-0997-1

Wardell JN, Stocks SM, Thomas CR, Bushell ME (2002) Decreasing the hyphal branching rate of *Saccharopolyspora erythraea* NRRL 2338 leads to increased resistance to breakage and increased antibiotic production. Biotechnol Bioeng 78:141–146. https://doi.org/10.1002/bit.10210

Warr GA, Veitch JA, Walsh AW, Hesler GA, Pirnik DM, Leet JE, Lin PF, Medina IA, McBrien KD, Forenza S, Clark JM, Lam KS (1996) BMS-182123, a fungal metabolite that inhibits the production of TNF-alpha by macrophages and monocytes. J Antibiot 49:234–240. https://doi.org/10.7164/antibiotics.49.234

Weisenberger S, Schumpe A (1996) Estimation of gas solubilities in salt solutions at temperatures from 273 K to 363 K. AIChE J 42:298–300. https://doi.org/10.1002/aic.690420130

Whitaker A (1992) Actinomycetes in submerged culture. Appl Biochem Biotechnol 32:23–35. https://doi.org/10.1007/BF02922146

Wilhelm E, Battino R, Wilcock RJ (1977) Low-pressure solubility of gases in liquid water. Chem Rev 77:219–262. https://doi.org/10.1021/cr60306a003

Willemse J, Büke F, van Dissel D, Grevink S, Claessen D, van Wezel GP (2018) SParticle, an algorithm for the analysis of filamentous microorganisms in submerged cultures. A Van Leeuw J Microb 111:171–182. https://doi.org/10.1007/s10482-017-0939-y

Wittler R, Baumgartl H, Lübbers DW, Schügerl K (1986) Investigations of oxygen transfer into *Penicillium chrysogenum* pellets by microprobe measurements. Biotechnol Bioeng 28:1024–1036. https://doi.org/10.1002/bit.260280713

Woo EJ, Starks CM, Carney JR, Arslanian R, Cadapan L, Zavala S, Licari P (2002) Migrastatin and a new compound, isomigrastatin, from *Streptomyces platensis*. J Antibiot 55:141–146. https://doi.org/10.7164/antibiotics.55.141

Wucherpfennig T, Hestler T, Krull R (2011) Morphology engineering - Osmolality and its effect on *Aspergillus niger* morphology and productivity. Microb Cell Fact 10:58. https://doi.org/10.1186/1475-2859-10-58

Wucherpfennig T, Kiep KA, Driouch H, Wittmann C, Krull R (2010) Morphology and rheology in filamentous cultivations. In: Laskin AI, Sariaslani S, Gadd GM (ed) Advances in Applied Microbiology. Elsevier, pp 89–136. https://doi.org/10.1016/S0065-2164(10)72004-9

Yang F-C, Liau C-B (1998) Effects of cultivating conditions on the mycelial growth of *Ganoderma lucidum* in submerged flask cultures. Bioprocess Eng 19:233–236. https://doi.org/10.1007/PL00009014

Yang YK, Morikawa M, Shimizu H, Shioya S, Suga K-I, Nihira T, Yamada Y (1996) Image analysis of mycelial morphology in virginiamycin production by batch culture of *Streptomyces virginiae*. J Ferment Bioeng 81:7–12. https://doi.org/10.1016/0922-338X(96)83111-4

Yatmaz E, Karahalil E, Germec M, Ilgin M, Turhan İ (2016) Controlling filamentous fungi morphology with microparticles to enhanced β-mannanase production. Bioprocess Biosyst Eng 39:1391–1399. https://doi.org/10.1007/s00449-016-1615-8

Yin P, Wang Y-H, Zhang S-L, Chu J, Zhuang Y-P, Chen N, Li X-F, Wu Y-B (2008) Effect of mycelial morphology on bioreactor performance and avermectin production of *Streptomyces avermitilis* in submerged cultivations. J Chin Inst Chem Eng 39:609–615. https://doi.org/10.1016/j.jcice.2008.04.008

Yu P-L, Dunn NW, Kim WS (2002) Lactate removal by anionic-exchange resin improves nisin production by *Lactococcus lactis*. Biotechnol Lett 24:59–64. https://doi.org/10.1023/A:1013865502420

Band 1 **Sunder, Matthias**: Oxidation grundwasserrelevanter Spurenverunreinigungen mit Ozon und Wasserstoffperoxid im Rohrreaktor. 1996. FIT-Verlag · Paderborn, ISBN 3-932252-00-4

Band 2 **Pack, Hubertus**: Schwermetalle in Abwasserströmen: Biosorption und Auswirkung auf eine schadstoffabbauende Bakterienkultur. 1996. FIT-Verlag · Paderborn, ISBN 3-932252-01-2

Band 3 **Brüggenthies, Antje**: Biologische Reinigung EDTA-haltiger Abwässer. 1996. FIT-Verlag · Paderborn, ISBN 3-932252-02-0

Band 4 **Liebelt, Uwe**: Anaerobe Teilstrombehandlung von Restflotten der Reaktivfärberei. 1997. FIT-Verlag · Paderborn, ISBN 3-932252-03-9

Band 5 **Mann, Volker G.**: Optimierung und Scale up eines Suspensionsreaktorverfahrens zur biologischen Reinigung feinkörniger, kontaminierter Böden. 1997. FIT-Verlag · Paderborn, ISBN 3-932252-04-7

Band 6 **Boll Marco**: Einsatz von Fuzzy-Control zur Regelung verfahrenstechnischer Prozesse. 1997. FIT-Verlag · Paderborn, ISBN 3-932252-06-3

Band 7 **Büscher, Klaus**: Bestimmung von mechanischen Beanspruchungen in Zweiphasenreaktoren. 1997. FIT-Verlag · Paderborn, ISBN 3-932252-07-1

Band 8 **Burghardt, Rudolf**: Alkalische Hydrolyse – Charakterisierung und Anwendung einer Aufschlußmethode für industrielle Belebtschlämme. 1998. FIT-Verlag · Paderborn, ISBN 3-932252-13-6

Band 9 **Hemmi, Martin**: Biologisch-chemische Behandlung von Färbereiabwässern in einem Sequencing Batch Process. 1999. FIT-Verlag · Paderborn, ISBN 3-932252-14-4

Band 10 **Dziallas, Holger**: Lokale Phasengehalte in zwei- und dreiphasig betriebenen Blasensäulenreaktoren. 2000. FIT-Verlag · Paderborn, ISBN 3-932252-15-2

Band 11 **Scheminski, Anke**: Teiloxidation von Faulschlämmen mit Ozon. 2001. FIT-Verlag · Paderborn, ISBN 3-932252-16-0

Band 12 **Mahnke, Eike Ulf**: Fluiddynamisch induzierte Partikelbeanspruchung in pneumatisch gerührten Mehrphasenreaktoren. 2002. FIT-Verlag · Paderborn, ISBN 3-932252-17-9

Band 13 **Michele, Volker**: CDF modeling and measurement of liquid flow structure and phase holdup in two- and three-phase bubble columns. 2002. FIT-Verlag · Paderborn, ISBN 3-932252-18-7

Band 14 **Wäsche, Stefan**: Einfluss der Wachstumsbedingungen auf Stoffübergang und Struktur von Biofilmsystemen. 2003. FIT-Verlag · Paderborn, ISBN 3-932252-19-5

Band 15 **Krull Rainer**: Produktionsintegrierte Behandlung industrieller Abwässer zur Schließung von Stoffkreisläufen. 2003. FIT-Verlag · Paderborn, ISBN 3-932252-20-9

Band 16 **Otto, Peter**: Entwicklung eines chemisch-biologischen Verfahrens zur Reinigung EDTA enthaltender Abwässer. 2003. FIT-Verlag · Paderborn, ISBN 3-932252-21-7

Band 17 **Horn, Harald**: Modellierung von Stoffumsatz und Stofftransport in Biofilmsystemen. 2003. FIT-Verlag · Paderborn, ISBN 3-932252-22-5

Band 18 **Mora Naranjo, Nelson**: Analyse und Modellierung anaerober Abbauprozesse in Deponien. 2004. FIT-Verlag · Paderborn, ISBN 3-932252-23-3

Band 19 **Döpkens, Eckart**: Abwasserbehandlung und Prozesswasserrecycling in der Textilindustrie. 2004. FIT-Verlag · Paderborn, ISBN 3-932252-24-1

Band 20 **Haarstrick, Andreas**: Modellierung millieugesteuerter biologischer Abbauprozesse in heterogenen problembelasteten Systemen. 2005. FIT-Verlag · Paderborn, ISBN 3-932252-27-6

Band 21 **Baaß, Anne-Christina**: Mikrobieller Abbau der Polyaminopolycarbonsäuren Propylendiamintetraacetat (PDTA) und Diethylentriaminpentaacetat (DTPA). 2004. FIT-Verlag · Paderborn, ISBN 3-932252-26-8

Band 22 **Staudt, Christian**: Entwicklung der Struktur von Biofilmen. 2006. FIT-Verlag · Paderborn, ISBN 3-932252-28-4

Band 23 **Pilz, Roman Daniel**: Partikelbeanspruchung in mehrphasig betriebenen Airlift-Reaktoren. 2006. FIT-Verlag · Paderborn, ISBN 3-932252-29-2

Band 24 **Schallenberg, Jörg**: Modellierung von zwei- und dreiphasigen Strömungen in Blasensäulenreaktoren. 2006. FIT-Verlag · Paderborn, ISBN 3-932252-30-6

Band 25 **Enß, Jan Hendrik**: Einfluss der Viskosität auf Blasensäulenströmungen. 2006. FIT-Verlag · Paderborn, ISBN 3-932252-31-4

Band 26 **Kelly, Sven**: Fluiddynamischer Einfluss auf die Morphogenese von Biopellets filamentöser Pilze. 2006. FIT-Verlag · Paderborn, ISBN 3-932252-32-2

Band 27 **Grimm, Luis Hermann**: Sporenaggregationsmodell für die submerse Kultivierung koagulativer Myzelbildner. 2006. FIT-Verlag · Paderborn, ISBN 3-932252-33-0

Band 28 **León Ohl, Andrés**: Wechselwirkungen von Stofftransport und Wachstum in Biofilsystemen. 2007. FIT-Verlag · Paderborn, ISBN 3-932252-34-9

Band 29 **Emmler, Markus**: Freisetzung von Glucoamylase in Kultivierungen mit *Aspergillus niger*. 2007. FIT-Verlag · Paderborn, ISBN 3-932252-35-7

Band 30 **Leonhäuser, Johannes**: Biotechnologische Verfahren zur Reinigung von quecksilberhaltigem Abwasser. 2007. FIT-Verlag · Paderborn, ISBN 3-932252-36-5

Band 31 **Jungebloud, Anke**: Untersuchung der Genexpression in *Aspergillus niger* mittels Echtzeit-PCR. 1996. FIT-Verlag · Paderborn, ISBN 978-3-932252-37-2

Band 32 **Hille, Andrea**: Stofftransport und Stoffumsatz in filamentösen Pilzpellets. 2008. FIT-Verlag · Paderborn, ISBN 978-3-932252-38-9

Band 33 **Fürch, Tobias**: Metabolic characterization of recombinant protein production in *Bacillus megaterium*. 2008. FIT-Verlag · Paderborn, ISBN 978-3-932252-39-6

Band 34 **Grote, Andreas Georg**: Datenbanksysteme und bioinformatische Werkzeuge zur Optimierung biotechnologischer Prozesse mit Pilzen. 2008. FIT-Verlag · Paderborn, ISBN 978-3-932252-40-120

Band 35 **Möhle, Roland Bernhard**: An Analytic-Synthetic Approach Combining Mathematical Modeling and Experiments – Towards an Understanding of Biofilm Systems. 2008. FIT-Verlag · Paderborn, ISBN 978-3-932252-41-9

Band 36 **Reichel, Thomas**: Modelle für die Beschreibung das Emissionsverhaltens von Siedlungsabfällen. 2008. FIT-Verlag · Paderborn, ISBN 978-3-932252-42-6

Band 37 **Schultheiss, Ellen**: Charakterisierung des Exopolysaccharids PS-EDIV von *Sphingomonas pituitosa*. 2008. FIT-Verlag · Paderborn, ISBN 978-3-932252-43-3

Band 38 **Dreger, Michael Andreas**: Produktion und Aufarbeitung des Exopolysaccharids PS-EDIV aus *Sphingomonas pituitosa*. 1996. FIT-Verlag · Paderborn, ISBN 978-3-932252-44-0

Band 39 **Wiebels, Cornelia**: A Novel Bubble Size Measuring Technique for High Bubble Density Flows. 2009. FIT-Verlag · Paderborn, ISBN 978-3-932252-45-7

Band 40 **Bohle, Kathrin**: Morphologie- und produktionsrelevante Gen- und Proteinexpression in submersen Kultivierungen von *Aspergillus niger*. 2009. FIT-Verlag · Paderborn, ISBN 978-3-932252-46-2

Band 41 **Fallet, Claas**: Reaktionstechnische Untersuchungen der mikrobiellen Stressantwort und ihrer biotechnologischen Anwendungen. 2009. FIT-Verlag · Paderborn, ISBN 978-3-932252-47-1

Band 42 **Vetter, Andreas**: Sequential Co-simulation as Method to Couple CFD and Biological Growth in a Yeast. 2009. FIT-Verlag · Paderborn, ISBN 978-3-932252-48-8

Band 43 **Jung, Thomas**: Einsatz chemischer Oxidationsverfahren zur Behandlung industrieller Abwässer. 2010. FIT-Verlag· Paderborn, ISBN 978-3-932252-49-5

Band 45 **Herrmann, Tim**: Transport von Proteinen in Partikeln der Hydrophoben Interaktions Chromatographie. 2010. FIT-Verlag · Paderborn, ISBN 978-3-932252-51-8

Band 46 **Becker, Judith**: Systems Metabolic Engineering of *Corynebacterium glutamicum* towards improved Lysine Prodction. 2010. Cuvillier-Verlag · Göttingen, ISBN 978-3-86955-426-6

Band 47 **Melzer, Guido**: Metabolic Network Analysis of the Cell Factory *Aspergillus niger*. 2010. Cuvillier-Verlag · Göttingen, ISBN 978-3-86955-456-3

Band 48 **Bolten J., Christoph**: Bio-based Production of L-Methionine in *Corynebacterium glutamicum*. 2010. Cuvillier-Verlag · Göttingen, ISBN 978-3-86955-486-0

Band 49 **Lüders, Svenja**: Prozess- und Proteomanalyse gestresster Mikroorganismen. 2010. Cuvillier-Verlag · Göttingen, ISBN 978-3-86955-435-8

Band 50 **Wittmann, Christoph**: Entwicklung und Einsatz neuer Tools zur metabolischen Netzwerkanalyse des industriellen Aminosäure-Produzenten *Corynebacterium glutamicum*. 2010. Cuvillier-Verlag · Göttingen, ISBN 978-3-86955-445-7

Band 51 **Edlich, Astrid**: Entwicklung eines Mikroreaktors als Screening-Instrument für biologische Prozesse. 2010. Cuvillier-Verlag · Göttingen, ISBN 978-3-86955-470-9

Band 52 **Hage, Kerstin**: Bioprozessoptimierung und Metabolomanalyse zur Proteinproduktion in *Bacillus licheniformis*. 2010. Cuvillier-Verlag · Göttingen, ISBN 978-3-86955-578-2

Band 53 **Kiep, Katina Andrea**: Einfluss von Kultivierungsparametern auf die Morphologie und Produktbildung von *Aspergillus niger*. 2010. Cuvillier-Verlag · Göttingen, ISBN 978-3-86955-632-1

Band 54 **Fischer, Nicole**: Experimental investigations on the influence of physico-chemical parameters on anaerobic degradation in MBT residual waste. 2011. Cuvillier-Verlag · Göttingen, ISBN 978-3-86955-679-6

Band 55 **Schädel, Friederike**: Stressantwort von Mikroorganismen. 2011. Cuvillier-Verlag · Göttingen, ISBN 978-3 86955-746-5

Band 56 **Wichter, Johannes**: Untersuchung der L-Cystein-Biosynthese in *Escherichia coli* mit Techniken der Metabolom- und ^{13}C-Stofffflussanalyse. 2011. Cuvillier-Verlag · Göttingen, ISBN 978-3-86955-750-2

Band 57 **Knappik, Irena Isabell**: Charakterisierung der biologischen und chemischen Reaktionsprozesse in Siedlungsabfällen. 2011. Cuvillier-Verlag · Göttingen, ISBN 978-3-86955-760-1

Band 58 **Driouch, Habib**: Systems biotechnology of recombinant protein production in *Aspergillus niger*. 2011. Cuvillier-Verlag Göttingen, ISBN 978-3-86955-808-0

Band 59 **Gehder, Matthias:** Development and Validation of Indicators for the Production and Quality of Seed Cultures. 2011. Cuvillier-Verlag Göttingen, ISBN 978-3-86955-847-9

Band 60 **Sommer, Becky:** Methodenentwicklung zur Charakterisierung sporenbildender Pilz-Seedingkulturen. 2011. Cuvillier-Verlag Göttingen, ISBN 978-3-86955-851-6

Band 61 **Dohnt, Katrin:** Charakterisierung von *Pseudomonas aeruginosa*-Biofilmen in einem *in vitro*-Harnwegskathetersystem. 2011. Cuvillier-Verlag Göttingen, ISBN 978-3-86955-852-3

Band 62 **Greis, Tillman:** Meddling the risk of chlorinated hydrocarbons in urban groundwater. 2011. Cuvillier-Verlag Göttingen, ISBN 978-3-86955-970-4

Band 63 **David, Florian:** Holistic bioprocess engineering of antibody fragment secreting *Bacillus megaterium*. 2012. Cuvillier-Verlag Göttingen, ISBN 978-3-95404-115-2

Band 64 **Palme, Wiebke:** Taxonomische Einordnung des Polyaminopolycarbonsäure-abbauenden Stammes BNC1 und Untersuchungen zum Abbau von 1,3 Propylendiamintetraacetat. 2012. Cuvillier-Verlag Göttingen, ISBN 978-3-95404-158-9

Band 65 **Lin, Pey-Jin:** Effect of fluid dynamics on pellet morphology and product formation of *Aspergillus niger*. 2012. Cuvillier-Verlag Göttingen, ISBN 978-3-95404-181-7

Band 66 **Kind, Stefanie:** Synthetic Metabolic Engineering of *Corynebacterium glutamicum* for Bio-based Production of 1,5-Diaminopentane. 2012. Cuvillier-Verlag Göttingen, ISBN 978-3-95404-264-7

Band 67 **Wilk, Franziska:** Charakterisierung der Stoffströme vorbehandelter Siedlungsabfälle in Deponiebioreaktoren. 2012. Cuvillier-Verlag Göttingen, ISBN 978-3-95404-281-4

Band 68 **Korneli, Claudia:** Target-oriented Bioprocess Optimization for Recombinant Protein Production in *Bacillus megaterium*. 2012. Cuvillier-Verlag Göttingen, ISBN 978-3-95404-289-0

Band 69 **Eslahpazir, Manely:** Numerical Characterization of Mechanical Stress and Flow Patterns in Stirred Tank Bioreactors. 2013. Cuvillier-Verlag Göttingen, ISBN 978-3-95404-449-8

Band 70 **Wucherpfennig, T.:** Cellular Morphology – A novel Process Parameter for the Cultivation of Eukaryotic Cells. 2013. Cuvillier-Verlag Göttingen, ISBN 978-3-95404-456-6

Band 71 **Buschke, Nele:** Bio-Nylon Monomers from Renewables using *Corynebacterium glutamicum*. 2013. Cuvillier-Verlag Göttingen, ISBN 978-3-95404-457-3

Band 72 **Bergmann, Sven:** Ectoine production by halotolerant microorganisms. 2013. Cuvillier-Verlag Göttingen, ISBN 978-3-95404-556-3

Band 73 **Hellriegel, Jan:** Engineering a Biofilm – Imitating Physico-Chemical Properties to improve Mechanical Characterization. 2014. Cuvillier-Verlag Göttingen, ISBN 978-3-95404-753-6

Band 74 **Berger, Antje:** Metabolische Netzwerkanalyse uropathogener *Pseudomonas aeruginosa*-Isolate. 2014. Cuvillier-Verlag Göttingen, ISBN 978-3-95404-762-8

Band 75 **Peterat, Gena:** Prozesstechnik und rekationskinetische Analysen in einem mehrphasigen Mikrobioreaktorsystem. 2014. Cuvillier-Verlag Göttingen, ISBN 978-3-95404-887-8

Band 76 **Godard, Thibault:** Systems biology of stress in *Bacillus megaterium* and its potential applications. 2016. Cuvillier-Verlag Göttingen, ISBN 978-3-7369-9336-5

Band 77 **Hönnscheidt, Christoph:** Entwicklung kolloiddisperser Wirkstoffformulierungen auf Basis von Biopolymeren. 2016. Cuvillier-Verlag Göttingen, ISBN 978-3-7369-9260-3

Band 78 **Walisko, Jana:** Morphologiebeeinflussung von *Lechevalieria aerocolonigenes* und heterologe Produktion von Rebeccamycin. 2017. Cuvillier-Verlag Göttingen, ISBN 978-3-7369-9502-4

Band 79 **Gädke, Johannes:** *In situ*-downstream processing of recombinant histidine-tagged proteins from cultivations of *Bacillus megaterium*. 2017. Cuvillier-Verlag Göttingen, ISBN 978-3-7369-9551-2

Band 80 **Lakowitz, Antonia:** Skalenübergreifende Produktion und Sekretion rekom--binanter Proteine mit Stämmen der Gattung *Bacillus*. 2017. Cuvillier-Verlag Göttingen, ISBN 978-3-7369-9576-5

Band 81 **Lladó Maldonado, Susanna Maria:** Bioengineering at the micro-scale: Design, characterization and validation of microbioreactors. 2019. Cuvillier-Verlag Göttingen, ISBN 978-3-7369-7025-0

Band 82 **Engel, Christina:** Quantitative analysis of the electrochemically active bacteria *Geobacter sulfurreducens* and *Shewanella oneidensis*. 2020. Cuvillier-Verlag Göttingen, ISBN 978-3-7369-7211-7

Band 83 **Tesche, Sebastian:** Production of labyrinthopeptin A1 with *Actinomadura namibiensis*. 2020. Cuvillier-Verlag Göttingen, ISBN 978-3-7369-7343-5

www.ingramcontent.com/pod-product-compliance
Ingram Content Group UK Ltd.
Pitfield, Milton Keynes, MK11 3LW, UK
UKHW022000190726
13853UKWH00004B/1642

9 783736 973503